Sida, Botanical Miscellany 17
Botanical Research Institute of Texas, Inc.

The Grasses of Barbados (Poaceae)

George Rogers and Ayanda Holder

Dept. of Biological and Chemical sciences
University of the West Indies
Cave Hill, barbados, West Indies

1999

ISSN 0833-1475
ISBN 1-889878-03-0

SIDA, BOTANICAL MISCELLANY NO. 17
FOUNDED BY
WM. F. MAHLER AND BARNEY L. LIPSCOMB, 1987
EDITOR: BARNEY L. LIPSCOMB
BOTANICAL RESEARCH INSTITUTE OF TEXAS, INC.
509 PECAN STREET
FORT WORTH, TEXAS 76102-4060 USA
CONTRIBUTING SPANISH EDITOR: FÉLIX LLAMAS
DPT. BIOLOGÍA VEGETAL, UNIVERSIDAD DE LEÓN
CAMPUS DE VEGAZANA S/N, 24071 LEÓN, SPAIN

COVER DESIGN/LINNY HEAGY

COVER PHOTO:
CHRIST CHURCH, RURAL AREA ONE QUARTER MILE
FROM LONG BEACH
PHOTO BY SOPHIA GREMMEL

FRONTISPIECE:
COIX LACRYMA-JOBI/JOB'S TEARS

THE GRASSES OF BARBADOS (POACEAE)
IS PUBLISHED WITH THE SUPPORT OF:
UNIVERSITY OF THE WEST INDIES
CAVE HILL, BARBADOS, WEST INDIES
BOTANICAL RESEARCH INSTITUTE OF TEXAS

DISTRIBUTION OF COPIES BY:

BOTANICAL RESEARCH INSTITUTE OF TEXAS
509 PECAN STREET
FORT WORTH, TEXAS 76102-4060
TELEPHONE: 1/817/332-4441
FAX: 1/817/332-4112
EMAIL: SIDA@BRIT.ORG

BRIT

SIDA, BOTANICAL MISCELLANY, NO. 17

PRINTED IN THE UNITED STATES OF AMERICA

The Grasses of Barbados

(POACEAE)

Table of Contents

List of Figures

Coix lacryma-jobi (Rogers 98-8)

THE GRASSES OF BARBADOS (POACEAE)

GEORGE ROGERS and AYANDA HOLDER

Dept. of Biological and Chemical Sciences
University of the West Indies
Cave Hill, Barbados, WEST INDIES

ABSTRACT

A floristic study of the grasses (Poaceae) of Barbados provides keys, descriptions, photographs, and specimen data for the 79 species in 45 genera growing free of active cultivation. The African *Dinebra retroflexa* had not been previously reported in the New World. Additional species added to the uncultivated Barbadian flora are *Digitaria bicornis* (Lam.) Roem. & Schult., *Echinochloa polystachya*, *Eragrostis acutiflora*, *Panicum diffusum*, *Sorghum halepense*, *Urochloa subquadripara*, and *Zoysia matrella*. Approximately half of the Barbadian grasses are probably introduced, and about half potentially indigenous. About 18 species are of conservation interest for being missing, or rare, or nuisances. A history of grass floristics in Barbados is provided, and remnant wild habitats are highlighted.

RESUMEN

En un estudio florístico de las gramíneas (Poaceae) de Barbados se ofrecen claves, descripciones, fotografías y datos de los especímenes de las 79 especies, pertenecientes a 45 géneros, que viven de modo natural sin cultivo activo. La especie africana *Dinebra retroflexa* no había sido citada previamente en el Nuevo Mundo. Especies adicionales que se añaden a la flora no cultivada de Barbados son *Digitaria bicornis* (Lam.) Roem. & Schult., *Echinochloa polystachya, Eragrostis acutiflora, Panicum diffusum, Sorghum halepense, Urochloa subquadripara*, y *Zoysia matrella*. Aproximadamente la mitad de las gramíneas de Barbados tienen probabilidad de ser introducidas, y cerca de la mitad son potencialmente naturales. Unas 18 especies tienen interés para la conservación por estar perdidas, ser raras o dañinas. Se ofrece una historia de la florística de las gramíneas en Barbados, y se resaltan los hábitats vestigiales naturales.

INTRODUCTION

Located a little north of 13 degrees N Latitude and a little west of 59 degrees W Longitude, Barbados is the easternmost island of the Lesser Antilles. The island is small (ca. 18 km X 26 km), rocky, and windswept. The eroded landmass is almost entirely derived from fossil coral. The varied topography is flat in some areas, rolling and hilly in others, sharply tiered in places, and in some areas cut by gullies. The highest elevation is about 340 m. The annual average rainfall is 100–200 cm, with some upland areas receiving more than this and some coastal areas receiving less. In addition to the limestone bedrock, the wetter uplands are mostly dominated by alkaline "red" (kaolinite) clay soils and the drier lowland sites are dominated by alkaline "black" (smectoid) clay soils. A wet season extends from approximately July to November and a pronounced dry season predominates from approximately January through May. There are about 700 species of wild plants, many of them are introduced.

Because sugarcane cultivation, pasturage, and disturbed scrubby/weedy vegetation cover most of the land, the more or less wild areas and specialized habitats are of particular interest. For instance, ecologically unusual are patchy areas to the north in the "Scotland District." In this region of dramatic topography and comparatively low disturbance occur rocky windswept hill-tops where the coral cap has eroded away to expose uniquely acid edaphic islands of sandstone (see Vernon & Carroll 1965). Such sites and neighboring lowlands where sandstone sediment has collected are the homes of some grasses, such as *Paspalum millegrana* and *P. virgatum*. One sandstone hilltop occurs within the most well known biological locality in Barbados: Turner's Hall Woods. In the heart of the Scotland District, Turner's Hall Woods is a remnant of minimally disturbed, mature, multitiered, deeply shaded, comparatively moist, tall forest and thus is the sole Barbados home to many species, for example *Eriochloa punctata, Panicum pilosum, Pharus lappulaceus,* and *Pharus latifolius.* A number of additional forest patches are scattered in sites unsuitable for agriculture, for residential development, or for tourism development. One of the larger remnants is on the East Coast in St. Joseph Parish—Hackleton's Cliff and the forest immediately beneath it, known in part as Foster's Hall Woods.

Still other forest remnants occur in the extensive system of moist, shaded, variably disturbed "gullies." These limestone ravines tend to be deep and steep-sided, and drain primarily to the west. Much of the island is surrounded by beaches, though a high percentage of these, in particular on the south and west coasts, are occupied by resort hotels. Those on the east coast remain more natural, including a long beach and dune area in and adjacent to a national park. The non-beach coastline is rocky and wetlands are limited. There exist freshwater springs, seeps, soggy areas, streams, and artificial ponds, and a handful of lovely fresh, clean, clear streams flow from the high center primarily to the east. Chancery Lane Swamp is an open, sunny, sometimes-dry brackish marsh behind the beach on the east coast in Christ Church Parish, and is the only place where *Paspalidium geminatum* was once collected. The largest wetland is the brackish Graeme Hall Swamp on the south coast in Christ Church Parish. Graeme Hall Swamp is a highly disturbed complex of mangrove swamp, marl, savannah, marsh, and weedy meadows. The swamp has long been manipulated and impacted for diverse purposes, and during the course of the present project has become home to a sewage treatment plant and to a landscaped tourist attraction.

Altogether 79 species in 45 genera were treated as growing outside of cultivation in Barbados. Because of their often enormous ranges, dispersal abilities, and frequent tendencies toward human-mediated dispersal, it is impossible to give "natural" ranges for many grasses. Beyond those dispersed to Barbados by ocean drift or birds are probably others that have come on the wind. Barbados receives windborne dust, insects, and probably fruits and seeds from Africa. Horticulture and agriculture have been practiced since pre-European times, with their diverse practitioners bringing crops and ornamentals, soil, and tagalong weeds. Livestock is abundant, and many of the dominant grasses are deliberately introduced fodder species, with others conceivably arriving in feeds or animal waste. Race horses are regular visitors. So are ships that discharge ballast and other debris. Lawns and golf courses abound, and escaped turf grasses are components of the uncultivated flora, most notably Bermuda Grass (*Cynodon dactylon*) and Zoysia (*Zoysia matrella*).

With many cases of "natural origin" impossible to settle, we estimate about half of the Barbadian grasses to come from the Old World (about 27 species) or to be pantropical (about 9 species). The majority of the known and suspected "Old World" and pantropical species are forage grasses and/or weedy. These include *Arundo donax*, *Bambusa vulgaris*, *Bothriochloa bladhii*, *B. pertusa*, *Brachiaria eruciformis*, *Cenchrus echinatus*, *Dactyloctenium aegyptium*, *Dichanthium annulatum*, *D. aristatum*, *Echinochloa colona*, *Eleusine indica*, *Panicum maximum*, *Pennisetum purpureum*, *Rhynchelytrum repens*, *Rottboellia cochinchinensis*, *Setaria barbata*, *Sorghum halepense*, *Urochloa mutica*, and *Vetiveria zizanioides.*

Approximately half of the "wild" grasses are reasonably regarded as probably indigenous to Barbados, or at least to the Caribbean region. Approximately 33 species range from the U.S. or from Mesoamerica to South America. *Chloris sagraeana*, *Lasiacis divaricata*, and *Paspalum laxum* are more or less limited naturally to the Caribbean. Most of the forest and gully species are probably indigenous to Barbados or to the greater Caribbean region, primarily *Eriochloa punctata*, *Ichnanthus pallens*, *Lasiacis divaricata*, *L. sorghoidea*, *Lithachne pauciflora*, *Olyra latifolia, Oplismenus hirtellus*, *Pharus lappulaceus*, and *P. latifolius*. The beach-dwelling species are comprised largely of probably native, though widespread, grasses: *Anthephora hermaphrodita*, *Eustachys petraea*, *Paspalum laxum*, *Paspalum vaginatum*, *Sporobolus indicus*, *S. coromandelianus*, and *S. virginicus*. Also often indigenous to Barbados or to the greater Caribbean are all or most of the *Chloris* species, which tend toward weediness, most of the *Paspalum* species, and an assortment of others, including such weeds as *Digitaria bicornis*, *D. insularis*, *Echinochloa polystachya*, *Leptochloa mucronata*, *L. virgata*, and *Urochloa fasciculata*. A non-taxonomic account of the natural history of the Barbadian grasses and sedges with comments on species of particular conservation interest is in Rogers (1998).

Species growing "wild" and included in the present treatment but not previously reported for Barbados are *Digitaria bicornis, Dinebra retroflexa, Echinochloa polystachya, Eragrostis acutiflora, Panicum diffusum, Sorghum halepense, Urochloa subquadripara,* and the turf species *Zoysia matrella*. Prominent cultivated ornamentals not previously documented for Barbados are *Pennisetum setaceum* (Forssk.) Chiov. and *Thysanolaena maxima* (Roxb.) Kuntze.

Realizing that most if not all of the following species are plentiful elsewhere, the following grasses are rare (or extirpated) and of conservation interest at a local level. (See also the list of

Doubtful and Excluded Species.) The list of rare/extirpated species is annotated with abbreviations: THW (limited to Turner's Hall Woods), M (missing = not documented more recently than publication of the *Flora of Barbados*, 1965), PN (probably native), and W (weed elsewhere). Grasses rare on Barbados are *Chloris sagraeana* (PN), *Echinochloa polystachya* (PN), *Eragrostis glomerata* (M, PN), *Eragrostis pectinacea* var. *miserrima* (M, PN), *Eriochloa punctata* (PN, THW), *Eustachys petraea* (PN), *Ichnanthus pallens* (PN), *Olyra latifolia* (PN, THW, Jack-in-the-Box Gully), *Panicum diffusum* (PN, W—encountered once, on the university campus), *Panicum laxum* (PN, THW), *Panicum pilosum* (introduced?, THW), *Paspalidium geminatum* (M, PN), *Paspalum millegrana* (PN), *Paspalum virgatum* (PN, W), *Pharus lappulaceus* (PN, THW), *Pharus latifolius* (PN, THW, less abundant than the preceding species), *Schizachyrium brevifolium* (M, pantropical with nativity unclear), and *Setaria parviflora* (PN, M).

In recent years the grasses of Barbados have fallen into botanical neglect. During the 1930s and early 1940s J.A. Allan surveyed the island for grasses and published an account with illustrations based on fresh materials. His efforts ended in 1943 but were not published until 1957. Although Allan did prepare herbarium specimens (mostly dated in the 1930s), his 1957 publication with few exceptions is based on fresh material and is not tied directly to specimens. Specimens are not cited. Duplicates of Allan's and other collections went to C. E. Hubbard at Kew, and many of these with Hubbard's annotations presently reside in the Barbados National Herbarium (BAR) at the University of the West Indies. Decay and insect damage to many is severe, however, and an unknown number disappeared decades ago. The BAR herbarium has been well-curated in recent decades. Allan's treatment has a set of fine line drawings. The treatment of grasses in the *Flora of Barbados* (Gooding et al., 1965) is primarily modified from Allan (1957), repeated essentially verbatim for long stretches, with improved keys and improved nomenclatural citations but with Allan's illustrations omitted. Coming largely from Allan (1957), the data in the *Flora of Barbados* treatment date back far earlier than the deceptively recent 1965 publication date. Thus there exists no floristic treatment of Barbados grasses compiled recently nor based directly on specimens. Gould (1979) treated the grasses of the Lesser Antilles, and the breadth of that work has been invaluable for identifications, for generic and species concepts, and for nomenclature. Gould's treatment, however, pays little direct attention to Barbados and involved no fieldwork here. Again, there are almost no specimen citations. Major recent floristic works highly pertinent to studying the grasses of Barbados, which were much consulted, include Davidse and Pohl (1994), Judziewicz (1991), McVaugh (1983), Pohl (1980), and Webster and Peterson (1996) as well as older Caribbean floras.

A confusing aspect of the older work must be pointed out. Allan and some of his predecessors and contemporaries (especially Bovell, McIntosh, and Gooding) labeled their collections with *species* numbers as opposed to the more conventional *collection* numbers familiar to all taxonomic botanists. In other words, materials in BAR bear apparent collection numbers that actually indicate earlier individuals' specific identifications. Thus the same numbers are applied to multiple collections. In the numerous situations where such specimens have had identifications corrected or have had names changed for other reasons, the old species numbers are now divided among several species. As examples of the sorts of situations encountered, in *Anthephora hermaphrodita* we came across specimens labeled *Gooding 312* (1940) as well as *Allan 312*. There are two collections bearing the number *Allan 312*: one made in 1936, the other in 1938. And so forth. The only defense from confusion is to be aware of this troublesome historical practice, and to include at least the dates when citing the older collections. We have conformed with standard modern practice by using collection numbers for recent collections.

The present project reflects approximately two years aimed at surveying the grasses of Barbados. The goal has been to account for every species of grass growing outside of intensive cultivation, with 1996–1999 field collections encompassing the vast majority. Adams (1972), Allan (1957), Amshoff and Henrard (1943), Correll and Correll (1982), Davidse and Pohl (1994), Fournet (1978), Gooding (1974), Gooding et al. (1965), Gould (1979), Hall (1978), Hitchcock (1936, 1950), Hitchcock and Chase (1917), Judziewicz (1991), McVaugh (1983), Pohl (1980), Proctor (1984), Sauget and Liogier (1974), Webster and Peterson (1996), and additional resources were often consulted. Nonetheless, all morphological data are newly derived from herbarium specimens collected in Barbados, except for the defining characteristics of genera. The generic characterizations and generic key have made extensive reference to Gould (1979), Pohl (1980), Clayton and Renvoize (1986), Watson and Dallwitz (1994), and a diverse assortment of other sources. GR examined non-Barbados materials of several species at the University of Florida Herbarium and at the University of Michigan Herbarium to help with difficult identifications and to clarify species concepts. Morphological terminology follows Hitchcock (1950). The collection numbers cited with the photographs refer to herbarium specimens collected from the same populations and physically close to the individuals photographed. There are instances where photographs and the cited collections did not originate on the same date, however.

Ornamental species but not crop species are included in the keys. Characters used in the keys apply to the plant groups only as they are encountered in Barbados. Using the key to genera depends on interpretation of spikelet morphology. This is difficult for genera whose spikelets are highly modified, so these appear early in the key identified by readily observed characters and reappear in places appropriate to their spikelet morphology. Keys to individual species are placed within the generic treatments.

KEY TO THE GENERA OF GRASSES IN BARBADOS

1. Spikelets with awns over 3 cm long .... **Themeda**
1. Spikelets awnless or with awns shorter than 3 cm 2
 2. Spikelets clustered into spinose burs .... **Cenchrus**
 2. Spikelets otherwise 3
  3. Spikelets in clusters of 4 in an urn-shaped involucre of connate glumes .... **Anthephora**
  3. Spikelets otherwise 4
   4. Spikelets with 1–many florets; when imperfect or sterile florets present, these generally above the perfect ones; spikelets laterally compressed (or not compressed); disarticulation generally above the glumes (various subfamilies, including giant reeds and bamboos) .... **Subkey A**
   4. Spikelets usually with 2 florets, the upper one perfect and the lower one staminate or sterile (often manifest as a sterile lemma); spikelets dorsally compressed; disarticulation generally below the glumes (Panicoideae) .... **Subkey B**

SUBKEY A

1. Plants massive reeds 2 or more m tall with large complex panicles (Arundinoideae) 2
 2. Panicle drooping, not feathery; cultivated, clump-forming reeds up to 1.5–3 m tall .... **Thysanolaena maxima** (TIGER GRASS)
 2. Panicle erect, feathery; uncultivated reeds of roadsides, drainage ditches, and stream beds 3
  3. Inflorescences bisexual; lemmas awned; leaves arranged more or less evenly along the culm .... **Arundo**
  3. Inflorescences unisexual; lemmas not awned; leaves in a distinctive fan-shaped arrangement .... **Gynerium**
1. Plants otherwise 4
 4. Leaf blades elliptic to (ob)lanceolate, the ligule membranous 5
  5. Large woody grasses; leaves with (often inconspicuous) cross-veins; spikelets bisexual .... **Bambusa**
  5. Forest floor grasses; leaves without cross-veins; spikelets unisexual 6
   6. Leaf blades with oblique veins, often over 3.5 cm wide, often oblanceolate; spikelets in pairs (Phareae) .... **Pharus**
   6. Leaf blades parallel-veined, seldom attaining 3.5 cm wide, usually elliptic to lanceolate; spikelets not paired (Olyreae) 7
    7. Pistillate florets restricted to the tips of axillary panicles or racemes; the indurate lemmas helmet-shaped .... **Lithachne**
    7. Pistillate florets in terminal panicles; the indurate lemmas ellipsoid .... **Olyra**
 4. Leaf blades linear or occasionally narrowly lanceolate, the ligule usually ciliate (infrequently membranous and then the leaf blades linear or nearly so) 8
  8. Inflorescence consisting of a single, vertical, fusiform spike up to ca. 2 cm long; leaf blades narrower than 2 mm .... **Zoysia**
  8. Inflorescence digitate, racemose, or paniculate; leaf blades narrower than or broader than 2 mm 9
   9. Inflorescence a panicle (primary branches rebranched) 10
    10. Spikelets with just one floret; lemmas uninerved (or with no nerves apparent) .... **Sporobolus**
    10. Spikelets multiflowered; lemmas trinerved .... **Eragrostis**
   9. Inflorescence racemose or digitate (primary branches not rebranched) 11

11. Inflorescence with numerous primary branches arranged along a central axis 12
11. Inflorescence with the primary branches digitate (in *Eleusine* often with one additional branch inserted on the axis below the others) 14
12. Lemmas awned; upper florets modified into a cluster of 3 awns .... **Bouteloua**
12. Lemmas awnless; upper florets variable but not modified into a single 3-awned organ 13
13. Inflorescence branches stiff and straight, shorter than 4 cm; glumes 5–6 mm long, including a long subulate awn; plants sprawling .... **Dinebra**
13. Inflorescence branches flexible and drooping, often over 4 cm long; glumes to approx. 2 mm long and awnless; plants upright .... **Leptochloa**
14. Spikelets usually 3 mm or longer, with multiple fertile florets 15
15. Spikes digitate with no extra spike inserted beneath the others, to ca. 3 cm long, the rachis extending beyond the spikelets as a bristle; spikelets ca. 3 mm long; glumes awned .... **Dactyloctenium**
15. Spikes digitate and often with an extra spike inserted alone below the others, often exceeding 3 cm long, the rachis not extending beyond the spikelets; spikelets often exceeding 3 mm long; glumes awnless .... **Eleusine**
14. Spikelets usually shorter than 3 mm, with just one fertile floret 16
16. Lemmas awned .... **Chloris**
16. Lemmas awnless 17
17. Second glume emarginate; leaf blades 8–24 cm long, wider than 2 mm; rachis broadened; beach grass .... **Eustachys**
17. Second glume with a pointed apex; leaf blades to 6.5 cm long, narrower than 2 mm; rachis narrow; ubiquitous lawn and weed grass .... **Cynodon**

SUBKEY B

1. Glumes soft-textured and resembling the texture of the lower lemma; fertile palea and lemma indurate (Paniceae) 2
2. Inflorescence in the form of a flat paddle with the spikelets more or less embedded .... **Stenotaphrum**
2. Inflorescence otherwise 3
3. Spikelets in spiny, burlike clumps .... **Cenchrus**
3. Spikelets otherwise 4
4. Spikelets in clusters of 4 and enclosed within an urceolate cuplike involucre made up of the connate first glumes .... **Anthephora**
4. Spikelets otherwise 5
5. Inflorescence a striking reddish or purplish plume .... **Rhynchelytrum**
5. Inflorescence otherwise 6
6. Inflorescences with bristles inserted beneath the spikelets 7
6. Inflorescences without bristles beneath the spikelets 8
7. Spikelets not disarticulating with bristles .... **Setaria**
7. Spikelets disarticulating with bristles .... **Pennisetum**
8. First glume present (sometimes very small); second glume and lower lemma variable, usually not equal, nor bivalvate 9
9. Spikelets usually paired with one (sub)sessile and the other pedicellate; second glume silky-pubescent .... **Digitaria**
9. Spikelets not paired with one (sub)sessile and the other pedicellate (*Ichnanthus* sometimes with unequally pedicellate pairs); second glume glabrous or pubescent (*Brachiaria eruciformis*) but not silky 10
10. Inflorescences panicles (or unbranched in *Schizachyrium*) 15
10. Inflorescences digitate or racemose (infrequently with a little secondary branching toward the base) 12
8. First glume absent; second (only) glume and lower lemma of equal size and forming a clamlike bivalvate cover over the fertile floret 11
11. Leaf blade tongue-shaped, that is, remaining broad far above the middle; spikelets elongate and pointed .... **Axonopus**
11. Leaf blade widest below or at middle, the upper portion attenuate; spikelets often suborbicular to ovate and often blunt-tipped .... **Paspalum**
12. Spikelets in 4 rows; both lemmas apiculate (or short-awned) apically .... **Echinochloa** [*E. colona* with the leaf blades eligulate.]
12. Spikelets otherwise 13
13. First glume 1/2 the length of the spikelet; grasses of wet habitats; inflorescence rachises ending in a point .... **Paspalidium**

13. First glume shorter than 1/3 the length of the spikelet; grasses of diverse habitats; inflorescence rachises not ending in a point. [First glume often oriented toward the rachis. Upper lemmas often with transverse ridges, but see also *Panicum maximum.*] 14
14. Spikelets densely pubescent .... **Brachiaria**
14. Spikelets glabrous or minimally pubescent .... **Urochloa**
15. Spikelets inserted obliquely .... **Lasiacis**
15. Spikelets straight 16
16. First glume awned (with awn ca. 1 cm long or longer) ...... **Oplismenus**
16. First glume awnless (sometimes short-mucronate) 17
17. Leaf blades often broader than 2 cm; creeping shade-loving grass; upper lemma winged at base .... **Ichnanthus**
17. Leaf blades usually narrower than 1.5 cm; habits and habitats diverse (*Panicum pilosum* is shade-loving but not creeping); upper lemma not winged at base 18
18. Spikelets in 1 or 2 rows; upper lemma apiculate or short-awned .... **Panicum**
18. Spikelets in 4 rows; upper lemma not apiculate or short-awned .... **Echinochloa**
[*E. colona* with the leaf blades eligulate.]

1. Glumes harder-textured than the lower lemma; fertile palea and lemma soft-textured. [Spikelets usually in pairs with one sessile and perfect and the other pedicellate and reduced.] (Andropogoneae) 19
19. Plants with more or less spherical white beads (6–11 mm diameter) in the inflorescence; leaf blades often exceeding 2.5 cm wide .... **Coix**
19. Plants without white beads in the inflorescence; leaf blades only rarely exceeding 2.5 cm wide 20
20. Spikelets with awns over 3.5 cm long; the longest glumes ca. 3 cm long; spikelets in fan shaped fascicles several cm long .... **Themeda**
20. Spikelets awnless or with awns shorter than 3 cm; glumes shorter than 3 cm long; spikelets not in fan-shaped fascicles 21
21. Inflorescence compact and pencil-shaped with the sessile spikelets embedded in cavities in the rachis, and with the pedicellate spikelets tightly appressed and on ribbon-shaped pedicels; plants hirsute with irritating hairs .... **Rottboellia**
21. Inflorescences otherwise; plants glabrous or pubescent but not hirsute with irritating hairs 22
22. First glumes with 1 or 2 abaxial pits or glands .... **Bothriochloa**
22. First glumes not pitted 23
23. Lower glume of sessile spikelet armed with spines; leaf blades sharply folded .... **Vetiveria**
23. Lower glume of sessile spikelet spineless; leaf blades flat or nearly so 24
24. Inflorescence recondite, ca. 1 cm long; leaf blades often with rounded apices .... **Schizachyrium**
24. Inflorescence exposed, well over 1 cm long; leaf blades mostly with pointed apices 25
25. Inflorescence consisting of 1–4 fingerlike spikes; nodes conspicuously bearded with erect hairs .... **Dichanthium**
25. Inflorescence paniculate; nodes glabrous or with appressed hairs 26
26. Spikelets awnless (fertile lemma of sessile spikelet sometimes with an acicular tip ca. 1 mm long); plants with lemon scent when fresh; inflorescence axes partly concealed by spathes .... **Cymbopogon**
[Cultivated, perhaps infrequently escaping.]
26. Sessile spikelets awned on fertile lemmas; fresh plants not lemon scented; inflorescence axes not partly concealed 27
27. Inflorescence axes strongly ascending and silky-pubescent; leaf blades to 7 mm wide .... **Andropogon**
27. Inflorescence axes more or less horizontal, glabrate or scabrous (or hirsute distally); leaf blades to 2 cm wide .... **Sorghum**

TAXONOMY

Andropogon L., Sp. Pl. 1045. 1753.

Inflorescences highly branched and paniculate. Spikelets paired, usually with one member of the pair sessile and bearing a perfect and a neuter floret (extremely reduced in *A. glomeratus*), the other member pedicellate and staminate or abortive (as in *A. glomeratus*). Spikelets disarticulating with rachis fragments, these sometimes bearing pedicels. Glumes lacking awns.

A large worldwide genus circumscribed differently by different authors. Segregate genera found in Barbados and sometimes included in *Andropogon* include *Bothriochloa*, *Cymbopogon* (cultivated), and *Schizachyrium*.

1. **Andropogon glomeratus** (Walter) B.S.P., Prelim. Cat. N.Y. 67. 1888. BASIONYM: *Cinna glomerata* Walter, Fl. Carol. 59. 1788.

Robust, coarse grass. Culms to 130 cm tall, to ca. 5 cm in diameter, smooth. Sheaths up to about 11 cm long, these arising above nodes represented by a pronounced dark ring (dried). Ligules 2 mm long, often conspicuous, white-translucent, erose-margined. Blades to 30 cm long and 7 mm wide, long-tapering, smooth. Inflorescence to 30 cm long and 9 cm wide, held well above leaves, dense and highly branched, with strongly ascending branches and branchlets; rachis zig-zag and densely silky-pubescent. Sessile spikelet with the neuter floret extremely reduced, the fertile floret with long, narrow glumes greatly exceeding the lemma, this delicate, hyaline, and bilobed, bearing a long awn inserted toward the base, the palea delicate. Pedicellate spikelet very reduced, awnless, its pedicel silky.

Field recognition characters.—Broomlike inflorescences having silky zig-zag axes.

Distribution and habitat.—*Andropogon glomeratus* ranges from the U.S. to the West Indies and to Central America. In Barbados this species grows on high, windswept rocky sites with thin soil. It occurs on Scotland District sandstone as well as on more coral-based rocky Barbados soils.

Representative specimens: **BARBADOS. St. Andrew:** Turner's Hall Woods, exposed slope, 16 Nov 1989, *Carrington 1104* (BAR); Turner's Hall Woods, marine deposits, 8 Apr 1937, *Stehle 381* (BAR). **St. Joseph:** Abundant on top of Hackleton's Cliff, 29 Nov 1996, *Rogers 96-141* (BAR, FLAS).

Anthephora Schreb., Beschr. Gr. 2:105. 1779.

Spikelets single-flowered (and with a sterile lemma), in clusters of 4 united by fusions among the first glumes, the entire cluster falling together.

A small mostly African genus with one New World species.

2. **Anthephora hermaphrodita** (L.) Kuntze, Revis. Gen. Pl. 2:759. 1891. (**Figs. 1, 2**). BASIONYM: *Tripsacum hermaphroditum* L., Syst. Nat. ed. 10, 2:1261. 1759.

Variable plants ranging from tufted and spreading to erect. Culms to 30(–60) cm tall and 1 mm in diameter. Sheaths 2–8 cm long, keeled. Ligules 1.5 mm long, composed of a tuft of off-white hairs. Blades up to 8(–14) cm long and 5(–7) mm wide, scabrous on the adaxial surface with minute hairs (concentrated on the margins). Inflorescences spicate, usually held well above the leaves, up to 6 cm long, with a strongly zig-zag axis. Spikelets 4–6 mm long and 2–3 mm wide, borne in urceolate 4-toothed involucres.

Field recognition characters.—Spikelets clustered in involucre resembling a tulip flower with four tepals. Spicate inflorescences zig-zag after dehiscence of the spikelet clusters.

Distribution and habitat.—*Anthephora hermaphrodita* ranges from Mexico to South America, including the West Indies. It is a common weed in Barbados, especially near the sea (Fig. 2).

Comments.—This species differs from other Barbados grasses by its distinctive clusters of four one-flowered spikelets united by the first glumes.

Representative specimens: **BARBADOS. Christ Church:** Seashore, Chancery Lane, Dec 1936, *Allan 312* (BAR). **St Andrew:** Along coast between Belleplaine and Barclays Park, 18 Sep 1996, *Rogers 96-59* (BAR, FLAS). **St. Lucy:** Seashore near Animal Flower Cave, Jan 1938, *Allan 312* (BAR); area above Animal Flower Cave, 19 Dec 1996, *Rogers 96-161* (BAR). **St. Philip:** Bel Air, dry pasture, Aug 1940, *Gooding 312* (BAR); East Pt. Lighthouse, area near cliffs, Oct 1996, *Forde 96-9* (BAR); near Crane Hotel [a beach resort], "6.10.1901," *Bovell 293* (BAR).

Arundo L., Sp. Pl. 81. 1753.

Enormous grasses with large feathery, highly branched inflorescences. Spikelets with numerous florets.

A small Old World genus.

3. **Arundo donax** L., Sp. Pl. 81. 1753.

Very large, coarse grass. Culms to 2 or more m tall, 12(–15) mm in diameter. Ligule a mem-

FIG. 1. *Anthephora hermaphrodita* (*Rogers 96-161*)
FIG. 2. Site for *Anthephora hermaphrodita*

brane 1.5 mm long. Blades up to 46 cm long and 2.7 cm wide, very narrowly lanceolate, drawn out to a long-attenuate apex, with a well-developed midnerve and conspicuous lateral nerves, the base sagittate. Inflorescences to 63 cm long and 12 cm wide, multi-branched, plumose, pubescent panicles, the branches ascending. Spikelets 12–14 mm long, on puberulent pedicels, multi-flowered. Glumes of nearly equal length, glabrous, very narrowly lanceolate. Lemmas long-lanceolate, bearing abundant long silky hairs. Paleas much shorter, blunt-tipped.

Field recognition characters.—Grass several m tall with sagittate leaf bases.

Common names.—Reed, Giant Reed.

Distribution and habitat.—*Arundo donax* is naturalized from the Old World. In Barbados it favors roadside ditches and stream bottoms.

Comments.—The similar *Gynerium sagittatum* has a distinctive fanlike leaf arrangement with blades to 6 cm wide, and inflorescences well over 1 m long (as opposed to leaves evenly spaced along the axis, inflorescences under 70 cm long, and leaf blades under 3 cm wide in *Arundo*). The locally cultivated Asian ornamental *Thysanolaena maxima* (Roxb.) Kuntze resembles *Arundo* but has drooping (vs. upright), non-feathery panicles with spikelets a fraction as long as those of *Arundo.*

Representative specimens: **BARBADOS.** Parish not specified. Fresh Water Club, sandy field near shore, Aug 1939, *Allan s.n.* (BAR). **St. Andrew:** Along Highway 2 near Greenland, by stream, 4 Sep 1996, *Rogers 96-28* (BAR, FLAS); Walkers, riverbank, Sep 1940, *Gooding 518* (BAR) [Presumed to be Walker's in St. Andrew Parish due to the identity of the specimen and due to the mention of a riverbank. There is a site known as "Walkers" in St. George Parish as well]. **St. James:** Weston, "9,7,1901," *Bovell 257* (BAR).

Axonopus P. Beauv., Ess. Agrostogr. 12. 1812.

Inflorescences with long thin spikes bearing solitary spikelets in 2 rows, the rachis broadened. Spikelets with 2 florets, the lower one staminate or sterile, the upper floret perfect. First glume wanting, the second glume and lemma of the lower floret of nearly equal length.

A worldwide genus of about 100 species, mostly neotropical.

4. **Axonopus compressus** (Sw.) P. Beauv., Ess. Agrostogr. 154. 1812. BASIONYM: *Milium compressum* Sw., Prodr. 24. 1788.

Spreading, often prostrate, stoloniferous grasses rooting at the nodes. Culms rising to ca. 40 cm tall, frequently with bearded nodes. Sheaths flattened, 2.5–15 cm long, glabrous except for occasional pubescence on the margins, especially near the apex. Blades 6–25(–31) cm long and 5–10 mm wide, rounded or acute (or acuminate) apically, with a well-developed midvein, glabrous except for the puberulent to hirsute margins. Inflorescences often subtended by a bract shorter than most of the foliage leaves, consisting of fascicles of 2–6 narrow fingerlike spikes 6–9.5 cm long borne along and/or topping a short main axis, robust plants with multiple fascicles per stem. Rachises flattened and slightly winged. Spikelets 2–2.2 mm long, glabrous or slightly silky-pubescent, the 2 rows borne on one side of the rachis. First glumes absent. Second glumes boat-shaped, wrapped marginally around spikelets. Lemmas of lower (sterile) florets almost as large as second glumes. Lemmas and paleas of upper (fertile) florets shorter, ovate, nearly equal in length to each other.

Field recognition characters.—Broad leaf blades, often with acute or rounded tips; thin wandlike spikes with flattened rachis axes.

Common name.—Carpet Grass.

Distribution and habitat.—A pantropical species with an obscure original range, *Axonopus compressus* is abundant in Barbados, where it occurs uncultivated in gullies, on forest floors, and on the seashore.

Comments.—This species is a popular lawn grass. *Axonopus compressus* might be confused in the field with *Digitaria bicornis*, which also has thin wandlike spikes with broadened rachises. *Axonopus compressus* differs from *Digitaria bicornis* by having tongue-shaped leaves staying broad until close to the apex (vs. leaves linear-lanceolate with long-attenuate apices), spikelets up to 2.2 mm long (vs. over 3 mm long), and narrower spike rachises (ca. 0.5 mm vs. ca. 1 mm wide).

Representative specimens: **BARBADOS. St. Andrew:** On dunes behind beach between Belleplaine and Barclays

Park, 18 Oct 1996, *Rogers 96-95* (BAR, FLAS); Turner's Hall Wood, moist undergrowth, Oct 1939, *Allan 15c* (BAR). **St. John:** Hackleton's Cliff, in the shade of the cliff, 4 May 1997, *Rogers 97-19* (BAR). **St. Joseph:** Andrews Plantation, Oct. 1935, *Mayers 15* (BAR); Rusher Gully, gully floor, Jun 1940, *Gooding 15c* (BAR). **St. Thomas:** Walkes Spring Gully, 21 Oct 1937, *McIntosh & Allan "446a = 15d"* (BAR).

Bambusa Schreb., Gen. Pl. 236. 1789, nom. cons.

Culms lignified and hard, often several meters tall, with multiple branches at nodes. Spikelets multi-flowered; glumes small; stamens 6.

A genus of some 70 species mostly of the Old World, none indigenous to Barbados.

KEY TO WOODY BAMBOOS (*BAMBUSA* AND *DENDROCALAMUS*) IN BARBADOS

1. Bases of culms often exceeding 12 cm in diameter; sheaths deciduous; leaf blades without cross veins, abaxially pubescent .... **Dendrocalamus strictus** (Roxb.) Nees
   [In Gooding et al. (1965), not encountered during the present study.]
1. Bases of culms up to ca. 12 cm in diameter; sheaths persistent; leaf blades with (often inconspicuous) cross veins, abaxially glabrous (*Bambusa*) 2
   2. Plants exceeding 3 m tall; leaf blades mostly wider than 1.8 cm .... **Bambusa vulgaris**
   2. Plants up to ca. 3 m tall; leaf blades narrower than 1.8 cm .... **Bambusa glaucescens** (Willd.) Merr.
      [*B. multiplex* (Lour.) Raeusch ex Schult. & Schult. f.) [cultivated]

5. Bambusa vulgaris Schrad. ex J. C. Wendl., Coll. Pl. 2:26, pl. 47. 1808.

Highly variable clump-forming bamboo. Culms to ca. 15 m tall with basal diameters to ca. 12 cm. Sheaths diverging at the apex after dehiscence of the blade, sometimes with a little apical-marginal pubescence. Pseudopetioles 2–5 mm long, auriculate, the auricles glabrous or marginally ciliate and sometimes with an apical bristle, with a raised ciliate collar around the abaxial base of the pseudopetiole. Ligules membranous, ca. 1 mm long. Blades to ca. 26 cm long and 3 cm wide, a little glaucous, glabrous except for the auricle region, lance-elliptic to narrowly oblong, having up to ca. 22 conspicuous longitudinal veins in addition to the midrib, serrulate at the margins, the apex long-attenuate, the base rounded to obtuse. Flowering stems in whorls, the pseudospikelets in clusters at well-separated nodes; pseudospikelets subtended by elliptic-ovate, apically ciliate scales. Florets multiple on an elongate articulate rachilla. Glume(s), if present, not distinguished from the scales at the bases of the pseudospikelets. Lemmas often over 1 cm long, lanceolate, apically ciliate, the margins wrapped around the paleas, these about as long as the lemmas but far narrower, 2-keeled, apically ciliate.

Field recognition characters.—This the only large bamboo known to us outside of active cultivation.

Distribution and habitat.—*Bambusa vulgaris* is a much-cultivated and naturalized Old World species abundantly escaped in Barbados.

Representative specimens: **BARBADOS. Christ Church:** In yard of seaside house along road connecting Miami Beach with Atlantic Shores, 3 Sep 1997, *Rogers 97-35* (BAR-sterile). **St. Joseph:** Large clump along road from Highway 3 to Sturges, across from Kingdom Hall of Jehovah's Witness, 12 m tall, stems to 10 cm in diameter, 1 Sep 1997, *Rogers 97-31* (BAR-sterile). **St. Michael:** Queens Park, cultivated, flowering, undated, *Bovell s.n.* (BAR); Prior Park [probably cultivated], flowering, 1903, *Botanic Station Herbarium 531* (BAR). **St. Thomas:** Welchman Hall Gully, 16 Dec 1997, *Rogers 97-67* (BAR-sterile).

Bothriochloa Kuntze, Revis. Gen. Pl. 2:762. 1891.

Inflorescence consisting of a central axis bearing nearly straight unbranched or minimally branched lateral branches. Spikelets in pairs, with one sessile and the other pedicellate, the latter staminate or sterile. Spikelets falling together attached to segment of rachis. The sessile spikelet with 2 florets, the lower one reduced and the upper one perfect. Lemma of the upper (perfect) floret of the sessile spikelet awned.

A small pantropical genus.

KEY TO *BOTHRIOCHLOA* SPECIES IN BARBADOS

1. Leaf blades usually hirsute; spikes per inflorescence usually 6 or fewer; main axis in inflorescence usually shorter than 1.5 cm .... **Bothriochloa pertusa**
1. Leaf blades usually glabrous (except for collar); spikes per inflorescence usually 8 or more; main axis in inflorescence usually longer than 2 cm .... **Bothriochloa bladhii**

6. **Bothriochloa bladhii** (Retz.) S.T. Blake, Proc. Roy. Soc. Queensland 80(6):62. 1969. BASIONYM: *Andropogon bladhii* Retz., Observ. Bot. 2:27. 1781.

Andropogon intermedius R. Br., Prodr. 1:202. 1810. [var. *acidulus* Stapf, Agric. News Barbados 15:179. 1916.]

Robust, meadow-forming, sprawling or upright grass. Culms sometimes rising to 2 m tall, the nodes glabrous or appressed-pubescent. Blades 7–20(or more) cm long, 2–5(–8) mm wide, the bases to 9 cm long, often with long silky hairs near the collar. Inflorescences 5–9 cm long, 2.5–5 cm wide, consisting of (8–)10–15(–20) unbranched or basally branched spikes to 6 cm long rising from a well-developed central axis 2.5–3 cm long. Pedicelled spikelet 3 mm long, sterile, the pedicel silky-villous. Sessile spikelet 3 mm long, the first glume often with an abaxial pit, hirsute, the upper (fertile) floret with a very reduced, hyaline, linear lemma bearing an awn 7–15 mm long, the corresponding palea reduced beyond recognition; lower (sterile) floret represented by a reduced hyaline lemma.

Field recognition characters.—Inflorescences fasciate, made up of numerous upright spikes along a well-developed central axis, with conspicuous pubescence and with long, geniculate awns. First glume with an abaxial pit (see also *B. pertusa*).

Common name.—Sour Grass.

Distribution and habitat.—*Bothriochloa bladhii* is a widespread Old World pasture species and is one of the more abundant species in Barbados, where it can be the dominant grass of roadsides and meadows.

Representative specimens: **BARBADOS. Christ Church:** Chancery Lane, swamp, 24 Jun 1990, *Carrington 1246* (BAR); Chancery Lane, meadow S of pond, 23 Nov 1996, *Rogers 96-136* (BAR). **St. George:** Dry fields, Jan 1940, *Allan 516* (BAR). **St. John:** Approx. 1 km S of St. Johns Church, in full sun on coral ledge between road and sugar field, covering the area like a lawn, 16 Aug 1996, *Rogers 96-7* (BAR, FLAS). **St. Philip:** Bottom Bay, on rocks overlooking the sea, 25 Sep 1996, *Rogers 96-67* (BAR).

7. **Bothriochloa pertusa** (L.) A. Camus, Ann. Soc. Linn. Lyon ser. 2, 76:164. 1931. BASIONYM: *Holcus pertusus* L., Mant. Pl. 2: 301. 1771. *Andropogon pertusus* (L.) Willd., Sp. Pl. 4:922. 1806.

Culms often sprawling and stoloniferous, and rooting at the nodes, these usually bearded with erect hairs. Sheaths 3–6 cm long, flattened. Blades 3.5–11(–16) cm long, 2–4 mm wide, tending to be hirsute on both surfaces. Inflorescences 4–7 cm long, 1.5–6 cm wide, consisting of (1–)2–6 soft, silky, flexuous, unbranched or basally branched spikes 2–5 cm long rising from a short central axis to 1.5(–2.5) cm long. Pedicellate spikelets 3–4 mm long, lanceolate, unarmed, the pedicel silky-pubescent. Sessile spikelet ca. 4 mm long, silky-pubescent around the insertion and a little on the first glume, this bearing a large, conspicuous mid-abaxial pit; second glume almost as long as the first glume, the upper (fertile) floret having an awn to ca. 8–19 mm long on the reduced, narrow lemma, the palea reduced beyond recognition. Lower floret represented by reduced, hyaline lemma.

Field recognition characters.—Leaf blades tending to be hirsute; first glume of sessile spikelet pitted (also true of *B. bladhii*); nodes bearded with erect hairs; inflorescences with fewer than 6 spikes on a short central axis.

Distribution and habitat.—*Bothriochloa pertusa* flourishes on seashore sites as well as inland in Barbados. It is indigenous to the Old World.

Comments.—*Bothriochloa pertusa* differs from *B. bladhii* by having smaller (mostly under 15 cm long vs. frequently exceeding 15 cm), usually hirsute leaf blades (vs. the blades mostly glabrous except for the region near the collar), fewer than 6 spikes per inflorescence (vs. usually over (8–)10), and a main axis in the inflorescence generally shorter than 1.5 cm (vs. usually exceeding 2 cm). The conspicuously bearded nodes resemble *Dichanthium* species, from which *B. pertusa* is readily separated by its more sharply pointed, pitted first glumes.

Representative specimens: **BARBADOS. Christ Church:** Between Miami Beach and Atlantic Shores, at edge of lawn on bluff overlooking the sea, Oct 1996, *Rogers 96-98* (BAR). **St. Andrew:** On dunes between Belleplaine and Barclays Park [seashore], 18 Oct 1996, *Rogers 96-97* (BAR). **St. George:** In broken aging pavement around oil pump, just N of Highway 5 intersection with road to Turnpike and Carmichael, 16 Oct 1996, *Rogers 96-82* (BAR, FLAS). **St. James:** West Coastal highway at Chapel Rd. #2, on beach, 29 Oct 1996, *Rogers 96-125* (BAR). **St. Lucy:**

Rocks above Animal Flower Cave [on bluff overlooking the sea], 19 Dec 1996, *Rogers 96-162* (BAR). **St. Philip:** Unmowed field on grounds of Beulah Methodist Church, the dominant grass, 25 Sep 1996, *Rogers 96-60* (BAR).

Bouteloua Lag. in Varied. Ci. (4):134. 1805, orth. cons.

Grasses with mostly basal leaves and narrow blades. Inflorescences racemose. Spikelets with one fertile floret below a variable number of sterile florets. Glumes usually lanceolate. Fertile lemmas trinerved, the fertile paleas two-nerved. Spikelets usually with awns borne variably on glumes or lemmas.

A small genus ranging from North America to Argentina.

8. **Bouteloua americana** (L.) Scribn., Proc. Acad. Nat. Sci. Philadelphia 1891:306. 1891. **(Figs. 3–5)**. BASIONYM: *Aristida americana* L., Syst. Nat. ed. 10, 2: 879. 1759.

Culms 40(–50) cm long, slender, tufted, often sprawling or procumbent. Sheaths 2–5 cm long, glabrous except for hairs at the junction with the blade. Ligules a band of white cilia ca. 1 mm long. Blades up to 18 cm long, 1–3.5(–6) mm wide, bearing scattered hairs up to 3 mm long (often with a pronounced tuft at the collar), the margins scabrous. Raceme 6–13 cm long, 1–4 cm wide, with up to ca. 8 uncrowded spicate branches 2–5 cm long inserted on one side of the axis, often 1–2 cm apart, each bearing 4–10 uncrowded spikelets. Spikelets 8–12 mm long, multi-awned. Glumes narrowly lanceolate, the first glume a little smaller than the second, scabrous-pubescent, short-awned, keeled. Lemma of lowest (fertile) floret narrowly lanceolate, with the midrib extending as an awn to 2.3 mm long, flanked by 2 short awns, the fertile palea longer than the blade of the lemma, not awned. Upper floret(s) reduced to a cluster of 3 awns up to 8 mm long.

Field recognition characters.—Loose one-sided raceme with horizontal branches, long spikelets (ca. 1 cm long) with several awns. Leaf blades often only 1 mm wide, ciliate.

Distribution and habitat.—*Bouteloua americana* is a distinctive clump-former on exposed limestone rocks. It is a species of Central America, South America, and the Caribbean.

Representative specimens: **BARBADOS.** No locality data, 1901, *Bovell 347* (BAR). **Christ Church:** South Point, in pockets of soil on rocks [in a weedy meadow] between the lighthouse and the sea, 20 Nov 1997, *Rogers 97-55* (BAR, FLAS). **St. Joseph:** Lawn, Andrews Factory, Nov 1935, *Mayers 9* (BAR). **St. Philip:** Three Houses Park, on limestone rocks, 6 Dec 1996, *Rogers 96-148* (BAR).

Brachiaria (Trin.) Griseb., Fl. Ross. 4:469. 1853.

As treated here, *Brachiaria* is a monotypic genus resembling *Panicum* and *Urochloa*. The circumscriptions of these genera are unsettled. Due to the recentness and comprehensive nature of their treatment, we have conformed with Davidse & Pohl (1994), treating species often included in *Brachiaria* under *Urochloa*. *Brachiaria eruciformis* is compared with *Urochloa* species under that genus.

A worldwide genus of some 100 species.

9. **Brachiaria eruciformis** (Sm.) Griseb., Fl. Ross. 4:469. 1853. BASIONYM: *Panicum eruciforme* Sm., Fl. Graec. Prodr. 1:40. 1806.

Culms rising to ca. 40 cm tall, sprawling, hirsute to glabrous, rooting at the nodes. Sheaths up to 4(–8) cm long, usually hirsute. Blades up to 7(–9) cm long and 1.5–5 mm wide, long-tapering to an attenuate apex, hirsute to glabrous, sometimes especially pubescent around the nodes. Inflorescences up to 8(–12) cm long and 3 cm wide, held well above the foliage, with (1–)3–8(–9) well-spaced upright lateral branches 0.8–4 cm long, these occasionally branched. Spikelets 2 mm long, in two rows along one side of the side branch, pedicellate, heavily pubescent. First glume reduced, oriented away from the rachilla; second glume large, conspicuous, pubescent. Lemma of lower floret similar to the second glume, with a small hyaline palea. Upper floret with indurate lemma wrapped around the palea.

Field recognition characters.—Pubescent weed with racemose-paniculate inflorescences having side branches up to 4 cm long and with the pubescent spikelets markedly to one side of the side-branches.

Distribution and habitat.—*Brachiaria eruciformis* is an abundant Old World weed in many habitats, including cultivated land.

FIG. 3. *Bouteloua americana* (on rocks above Pearl Ave., Christ Church)
FIG. 4. *Bouteloua americana* (same locality)

Representative specimens: **BARBADOS.** No data, *Herbarium, Dept. of Agriculture, Barbados # 448* (BAR). **St. John:** Lodge School, edge of garden path, Dec 1940, *Gooding 21* (BAR). **St. Joseph:** Andrews Plantation, hedgerow, Sep 1936, *Mayers 21a* (BAR); Andrews Plantation, pasture, Nov 1935, *Mayers 21* (BAR). **St. Michael:** Codrington House, "24.9.01," *Bovell 448* (BAR). **St. Peter:** Along Highway 1 near Portland Plantation, weed in bean field, 4 Sep 1996, *Rogers 96-27* (BAR, FLAS). **St. Philip:** Sugar field weed at Three Houses, 6 Dec 1996, *Rogers 96-149* (BAR, FLAS).

Cenchrus L., Sp. Pl. 1049. 1753.

Sheaths compressed and the blades flat. Inflorescences spicate. Spikelets wrapped in spiny burs.

A small worldwide genus.

10. Cenchrus echinatus L., Sp. Pl. 1050. 1753. (**Fig.** 6).

Culms to 50 cm tall, 1 mm in diameter at the base of the inflorescence, sprawling to erect, tufted, with darkened nodes. Sheaths 4.5–6.5 cm long, silky-pubescent to glabrous. Blades to 14 cm long, 3–5(–6) mm wide, often crisped along the margins, the tips long-attenuate, adaxially usually with variable amounts of long silky pubescence or merely puberulent, the abaxial surface glabrous or nearly so. Inflorescence vertical and spikelike (though with short concealed second-order branches), 3–4 cm long, 12–15 mm wide, scabrous to hirsute on axes and on spikelet clusters. Spikelet clusters broad and basally flat, sharply spinose, 6–8(–10) mm wide and 6–8 mm tall, with the outer, narrow spines mostly shorter than the inner, flat spines.

Field recognition characters.—Unique spinose bur.

Common name.—Sand Spur.

Distribution and habitat.—*Cenchrus echinatus* extends from the U.S. to the Caribbean and South America, and is naturalized in Australia and in the Old World.

FIG. 5. *Bouteloua americana* (same locality)
FIG. 6. *Cenchrus echinatus* (*Rogers 96-160*)

Comments.—Gould (1979) listed *Cenchrus brownii* Roem. & Schult. for Barbados with no specimen cited. We have failed to find *C. brownii* in the Barbados National Herbarium, nor in the field.

Representative specimens: **BARBADOS.** No locality data. Weed, May 1901, *Freeman 171* (BAR). **Christ Church:** Roadside along ABC Highway near Kingsland, 11 Sep 1996, *Rogers 96-47* (BAR). **St. George:** Gun Hill, dry waste ground, Dec 1938, *Allan 470a* (BAR). **St. Joseph:** Andrews Sugar Factory, lawn, Dec 1935, *Mayers 470* (BAR). **St. Lucy:** Seashore, rocks above Animal Flower Cave, 19 Dec 1996, *Rogers 96-160* (BAR).

Chloris Sw., Prodr. 25. 1788.

Grasses with digitate (in ours) spikes having the spikelets arranged in two rows on one side of the rachis. Spikelets with one perfect floret below one or more reduced florets. Glumes unequal. Lemma of the perfect floret trinerved, pubescent, long-awned (thereby differing from the similar genus *Eustachys*).

A genus of about 50 species in warm climates around the globe.

KEY TO *CHLORIS* AND *EUSTACHYS* SPECIES IN BARBADOS

1. Leaf margins smooth except for serrulate to hispidulous apical margins; spikelets standing at right angles to the rachis, awnless (except for cusp ca. 0.3 mm long on the second glume) .... **Eustachys petraea**
1. Leaf margins scabrous or hispidulous for their entire length; spikelets appressed to rachis, awned from the lemmas 2
   2. Spikelets 3-awned .... **Chloris barbata**
   2. Spikelets 2-awned 3
      3. Longest awns under 2 mm long; spikes partly hidden by upper leaf sheath; fertile lemmas with dense, crowded cilia ca. 1 mm long on margins and midrib .... **Chloris ciliata**
      3. Longest awns 6 mm long or longer; spikes held well above foliage; fertile lemmas glabrous or with pubescence otherwise 4
         4. Sterile spikelet ca. 0.5 mm long; longest awns on the spikelets ca. 12–18 mm long...... **Chloris radiata**
         4. Sterile spikelet ca. 1.5 mm long; longest awns on the spikelets 6–12 mm long .... **Chloris sagraeana**

11. **Chloris barbata** (L.) Sw., Fl. Ind. Occid. 1:200. 1797 [as a nom. nov.]. (**Figs. 7, 8**). BASIONYM: *Andropogon barbatus* L., Syst. Nat. ed. 10, 2:1305. 1759.

Chloris inflata Link, Enum. Hort. Berol. Alt. 1:105. 1821.

Culms to ca. 70 cm(–1 m) tall, ca.1 mm in diameter at the apex, upright, the nodes darkened. Sheaths to 13.5 cm long, glabrous. Blades 4–20(–26) cm long and 1–4 mm wide, minutely serrulate marginally, abaxially glabrous or less often glabrate, adaxially usually puberulent to occasionally glabrous or glabrate. Inflorescences digitate with 6–10 spikes 5–8 cm long. Spikelets (excluding awns) ca. 2.5 mm long, in 2 rows along a winged rachis, dehiscing above the glumes, these infused with purple, glabrous, lanceolate to lance-elliptic, the first glume much shorter than the second glume. Lemma of perfect floret obovate, with a strongly developed midvein, silky-pubescent, tipped with a purplish awn ca. 6 mm long. Above this the spikelet with 2 sterile, reduced, awned lemmas.

Field recognition characters.—Digitate inflorescences with spikes bearded by having three purplish awns per spikelet.

Common name.—Finger Grass.

Distribution and habitat.—*Chloris barbata* is pantropical. It is one of the most pervasive and conspicuous weedy grasses in Barbados, being ubiquitous in disturbed places and sometimes forming extensive stands.

Representative specimens: **BARBADOS. Christ Church:** Graeme Hall Swamp, in *Pluchea* stand, damp swampy area by drainage ditch, "80/02/08," *Riven-Ramsey 31* (BAR); edge of drain in *Pluchea* stand, area filled with dredging, Graeme Hall Swamp, "80/01/23," *Riven-Ramsey 22* (BAR). **St. Joseph:** Weed at Andromeda Botanic Gardens, 27 Jan 1998, *Rogers 98-7* (BAR). **St. Michael:** Lower Estate, roadside trench, weed, Nov 1935, *Mayers 17* (BAR). **St. Philip:** Canefield at base of Mt. Pleasant, 25 Sep 1996, *Rogers 96-62* (BAR, FLAS); Station House Hill, roadside ditch, Jul 1940, *Gooding 17* (BAR).

12. **Chloris ciliata** Sw., Prodr. 25. 1788.

Culms sometimes exceeding 40 cm tall, with darkened nodes. Sheaths 5.5–10.5 cm long. Blades to 27 cm long, 1–6 mm wide. Inflorescences with ca. 5 digitate spikes 5–7 cm long, these covered basally by upper sheaths, the rachis hispid. Spikelets ca. 2.3 mm long (excluding awns). Glumes lance-elliptic, persistent on the rachis, scabrous on the strongly developed midrib. Fertile lemmas broadly ovate when unfolded, bearing an awn ca. 1.8 mm long between two small cusps, with dense straight cilia ca. 1 mm long on the margins and midrib. Sterile lemmas 2, broadest near the apex, with broad hyaline margins, glabrous, the larger one short-awned.

Field recognition characters.—Spikes ca. 5, partially hidden by upper leaf sheath, the rachis hispid; spikelets 2-awned with the longer awns under 2 mm long; fertile lemma with long cilia.

Distribution and habitat.—In Barbados, this species is known only from the southern coast in Christ Church Parish. It ranges from the southern U.S. to Argentina.

Representative specimens: **BARBADOS. Christ Church:** South Point Lighthouse Road, sandy field, dry soils, weed, Feb 1937, *Allan 514* (BAR); Worthing, sandy field, light soil, Mar 1940, *Allan 514a* (BAR); Oistins, along stream (but not in the water) in wet meadow across the Coast Road from the Oistins Post Office, 5 Sep 1997, *Rogers 97-37* (BAR).

13. **Chloris radiata** (L.) Sw. Prodr. 26. 1788. BASIONYM: *Agrostis radiata* L., Syst. Nat. ed. 10, 2:873. 1759.

Culms to 70 cm tall. Sheaths hirsute (especially apically and marginally) to glabrous. Blades highly variable, 3–(rarely)22 cm long and 6 mm wide, hirsute to glabrous on both surfaces, the margins scabrid. Inflorescences racemose with (4–)8–12(–17) flexuous spikes 4–8 cm long, the rachis hispidulous. First glume much shorter than the second glume, both lanceolate-linear, scabrous-hispidulous on the midrib. Fertile lemma lanceolate, ca. 3 mm long, with 2 acuminate cusps flanking the awn, this 12–18 mm long. Sterile lemma ca. 0.5 mm long, with an awn shorter than that on the fertile lemma.

Field recognition characters.—See below.

Distribution and habitat.—*Chloris radiata* ranges from Mexico to South America. In Barbados it is scattered and weedy.

FIG. 7. *Chloris barbata* (*Rogers 98-7*)
FIG. 8. *Chloris barbata* (*Forde 96-18*)

Comments.—*Chloris radiata*, *C. barbata* and *C. sagraeana* are all superficially similar. *Chloris barbata* differs from the other two by having three florets and three awns per spikelet as opposed to two florets and two awns. It has the most pronounced purplish coloration in the living inflorescence and is the most abundant of the three. *Chloris radiata* differs from *C. sagraeana* by having a much smaller sterile spikelet (ca. 0.5 mm long vs. 1.5 mm long), and the longest awns ca. 12–18 mm (vs. 6–12 mm) long. Seen *in situ*, *Chloris radiata* has a more delicate appearance. *Chloris radiata* tends inconsistently toward hirsute foliage. The statement in Gooding et al. that *C. radiata* differs from *C. sagraeana* by having over 10 spikes that stand upright even at maturity is inaccurate, since individuals of *C. radiata* often have as few as four spikes, and the spikes are often reflexed at maturity.

Representative specimens: **BARBADOS.** No locality data. "10.6.01," and "10.6.02," *Bovell 266* (3 specimens BAR). **Christ Church:** Maxwell Coast, foreshore, dry ground, Jan 1937, *Allan 317b* (BAR); roadside weed between Oistins and Newton, 10 Sep 1997, *Rogers 97-40* (BAR). **St. George:** In broken aging pavement around oil pump, just N of Highway 5 intersection with road to Turnpike and Carmichael, 16 Oct 1996, *Rogers 96-83* (BAR, FLAS); Gun Hill, Rose Cottage, usually fairly dry, weed, Jan 1938, *Allan 317a* (BAR). **St. Joseph:** Meadow on top of Hackleton's Cliff, 29 Nov 1996, *Rogers 96-143* (BAR); Industry, roadside ditch, common wayside grass, Aug 1940, *Gooding 317* (BAR).

14. **Chloris sagraeana** A. Rich., Hist. Fis. Cuba, Bot. XI: 315. 1850.

Culms to ca. 65 cm tall. Sheaths up to ca. 12 cm long, flattened and equitant. Blades 3–30 cm long, up to ca. 2.5 mm wide, linear and very narrow, glabrous or puberulent on both surfaces, the margins scabrous. Inflorescences digitate (or with a short central axis), bearing 3–6 spikes 4.8–8.5 cm long, these widely divergent at maturity. Spikelets 2.8–3 mm long (excluding the awns). Glumes shorter than the spikelet, the first glume shorter than the second glume, both lanceolate with long-attenuate apices. Fertile lemma sharply folded, tipped with an awn ca. 6–12 mm long, this flanked by 2 cusps, silky pubescent. Sterile lemma ca. 1.5 mm long, oblanceolate-elliptic (when unfolded), bearing between 2 cusps an awn slightly shorter than that of the fertile lemma.

Field recognition characters.—Narrow linear blades with equitant bases; spikes widely divergent when mature; spikelets with 2 awns.

Distribution and habitat.—To date, all Barbados localities are on the central East Coast near the sea. The overall range is limited to the Antilles.

Comments.—*Chloris sagraeana* could be confused with *C. radiata* but has strerile spikelets about 1.5 mm long as opposed to a fraction of that length.

Representative specimens: **BARBADOS.** No data. *Barbados Museum and Historical Society 506* (BAR). **St. John:** Consett's Bay, clay area, near sea shore, Jul 1940, *Gooding 517* (BAR). **St. John/St. Joseph:** Windward coast/Bath-Bathsheba, dry sandy field behind seashore, weedy but grazed by animals, Dec 1940, *Allan 506* (BAR). **St. Philip:** On rocky slope leading down to the sea from the East Point Lighthouse, sunny, exposed to salt spray, scarce, 9 Oct 1997, *Rogers 97-45* (BAR, FLAS).

Coix L., Sp. Pl. 972. 1753.

Large robust, upright grasses. Peduncles bearing a hard, white bead beneath the inflorescence. Inflorescences imperfect, the pistillate inflorescence held within the bead; staminate (or sterile) spikelets borne distal to the bead.

A small (sometimes interpreted as just one species) Old World genus spread through the tropics for fodder and for the decorative beads, which in some strains are eaten.

15. **Coix lacryma-jobi** L., Sp. Pl. 972. 1753. (**Fig. 9**).

Tall, branched glabrous grass with inflorescences and peduncles subtended by the upper leaves. Culms 1–3 mm in diameter toward the apex, the nodes darkened. Sheaths 2–6.5 cm long, striate with conspicuous veins. Blades lanceolate, 4–30 cm long, 20–34 mm wide, long-acuminate apically with a conspicuous midvein and several conspicuous lateral nerves, serrulate marginally, the base auriculate. Axillary peduncles 1–10 cm long, solitary or clustered, bearing 1(–2) glossy white globose to elliptic-ovate bead(s) 6–11 mm in diameter (or in the longer dimension when elongate). Pistillate inflorescences contained within the beads and having styles protruding 12–30 mm from an opening at the apex of the beads (pistillate spikelets occasionally produced apart from the beads). Staminate inflorescences protruding 2–4 cm from the apical opening. Staminate spikelets 9–11 mm long, usually in clusters of three with one pedicellate and

FIG. 9. *Coix lacryma-jobi* (*Rogers 98-8*)

two sessile (or paired with the pedicellate spikelet vestigial), with two florets, these 3-staminate, one usually larger than the other, the smaller one sometimes lacking stamens and with apparent vestigial gynoecium; glumes nearly equal, as long as the spikelet, a little scabrous on the midrib, elliptic, the first glume with two adaxial thickened lines converging apically and flanked by hyaline margins; fertile lemma lanceolate, about as long as the spikelet, the palea almost as large as the lemma.

Field recognition characters.—Leaves very wide, with auriculate bases; inflorescences with white beads.

Common name.—Job's Tears.

Distribution and habitat.—*Coix lacryma-jobi* is an Asian species occasionally seen persisting from cultivation in meadows and near settled areas.

Representative specimens: **BARBADOS.** No locality data or date. *Botanic Station Herbarium, Barbados No. 568.* **St. Andrew:** Along road to St. Simons coming from Haggats, 18 Sep 1996, *Rogers 96-56* (BAR); Spring Vale, roadside ditch, Oct 1940, *Gooding 318* (BAR); Turner's Hall Woods, meadow, 27 Jan 1998, *Rogers 98-8* (BAR). **St. Joseph:** Hills above Bathsheba, edge of canefield (fairly moist), cultivated sometimes for "beads," Jan 1934, *Allan 318* (BAR). **St. Michael:** Bank Hall, 30 Oct 1919, *Herbarium Dept. of Agriculture, Barbados No. 568* (BAR).

Cymbopogon Spreng., Pl. Pugill. 2:14. 1815.

Racemes paired and subtended by spathes. Spikelets paired with one sessile and perfect, and the other pedicellate and staminate or sterile, or spikelets in triads with the sessile spikelet flanked by two pedicellate spikelets. The lower spikelet pair of one or both racemes sterile.

A genus of about 60 Old World warm-climate species.

KEY TO *CYMBOPOGON* SPECIES IN BARBADOS

1. Fertile lemma with an awn 6–7 mm long; blades often wider than 1 cm ...... **Cymbopogon nardus** (L.) Rendle [May be encountered cultivated. Specimen in BAR: Dodds (St. Philip Parish), 26 Sep 1901, *Bovell 265*.]
1. Fertile lemma awnless or with an acicular tip ca. 1 mm long; blades narrower than 0.8 cm .... **Cymbopogon citratus** [Restricted to cultivation or nearly so.]

16. **Cymbopogon citratus** (DC.) Stapf, Bull. Misc. Inform. Kew 1906:322, 357. 1906. BASIONYM: *Andropogon citratus* DC., Cat. Pl. Horti Monsp. 78. 1813.

Large, erect, clump-forming, glaucous grass. Culms to 4 mm in diameter. Sheaths to 11 cm long. Blades to 44 cm long and 8 mm wide. Inflorescences up to 24 cm long and 5 cm wide, paniculate, the primary branches with zig-zag rachises; ultimate axes partly covered with lanceolate, folded spathes 1.7–4 cm long. Spikelets in pairs, with one member of the pair sessile and the other pedicellate. Pedicellate spikelets ca. 5 mm long, on densely hirsute pedicels, otherwise glabrous, the first glume as long as the spikelet, lance-elliptic, multinerved, wrapped around the second glume, this the same shape and almost as large as the first glume, wrapped around the lemma; lemma lance-elliptic, shorter than the glumes, enclosing a staminate flower; palea not apparent. Sessile spikelets ca. 5 mm long; first glume lance-elliptic, as long as the spikelet, 2-keeled; second glume a little shorter than the first glume, 1-keeled with a central mid-nerve, lance-elliptic; sterile lemma hyaline, delicate, much smaller than the glumes; fertile lemma 3/4 the length of the glumes, scarcely awned (drawn out into an acicular tip ca. 1 mm long).

Field recognition characters.—Cultivated glaucous grass with lemon scent; leaves erect, under 8 mm wide; panicle having branchlets in spathes; flowers awnless or with short awns ca. 1 mm long.

Common name.—Lemon Grass.

Comments.—*Cymbopogon citratus* is cultivated in Barbadian gardens for its lemon-scented foliage, which makes a pleasing tea, and is used otherwise for flavoring. This introduced species is not known to us outside of cultivation but is included, because it may be encountered escaped or persisting from prior cultivation. Our description is based on a single specimen.

Representative specimen: **BARBADOS.** No locality data. Cultivated for culinary purposes, Mar 1941, *Allan 518* (BAR).

Cynodon Rich., Syn. Pl. 1:85. 1805, nom. cons.

Inflorescence made up of digitate spikes bearing spikelets in 2 rows on a slender (vs. broadened as in *Digitaria*) rachis. Spikelets with one perfect floret (sometimes with a vestigial floret above this).

A small essentially Old World genus of much-branched spreading grasses.

17. **Cynodon dactylon** (L.) Pers., Syn. Pl. 1:85. 1805. BASIONYM: *Panicum dactylon* L., Sp. Pl. 58. 1753.

Aggressive creeping nearly glabrous grass, with prolific stolons. Culms rising to 32 cm tall. Sheaths to 6.5 mm long, often ciliate at the collar. Blades up to 6.5 cm long and up to 2 mm wide, narrow and uncrowded. Inflorescences with 3–5 digitate spikes 1.2–8 cm long, the rachis scabrous-puberulent on the angles. Spikelets ca. 2–2.5 mm long, in 2 rows on underside of spike. Glumes shorter than the spikelet, lanceolate, single-nerved, the first glume a little shorter than the second glume, with one floret. Lemma sharply creased, ovate-lanceolate (while folded), puberulent on the midvein.

Field recognition characters.—Leaves acicular and under 2 mm wide; inflorescences digitate and lacking awns (thereby similar to some *Digitaria* species but readily distinguished in the field by narrow leaves, small stature, non-flattened rachis, and spikelets not in pairs with one on a long pedicel and the other short-pedicellate or sessile.)

Common names.—Bermuda Grass, Devil Grass

Distribution and habitat.—*Cynodon dactylon* occurs around the globe in warm climates as a turf species with multiple cultivars and as a weedy nuisance. It is abundant throughout Barbados.

Comments.—*Axonopus compressus* can have somewhat similar inflorescences but differs by having winged rachises and far broader leaf blades.

Representative specimens: **BARBADOS.** No locality data. "13.5.1901," *Freeman & Bovell 378* (BAR). **St. Andrew:** Beach, S of Belleplaine and N of Barclays Park, 18 Oct 1996, *Rogers 96-93* (BAR). **St. John:** Lodge School, June 1940, *Gooding 469* (BAR). **St. Joseph:** Andrews Factory, lawn, Dec 1935, *Mayers 469* (BAR).

Dactyloctenium Willd., Enum. Pl. 1029. 1809.

Spikes sessile, digitate, with laterally flattened spikelets in 2 rows. Glumes both well-developed, with a single midrib. Palea about as long as the lemma, this trinerved. Fruit subglobose, the seeds wrinkled and enclosed in a thin pericarp.

About four species of the Old World.

18. Dactyloctenium aegyptium (L.) Willd., Enum. Pl. 1029. 1809. (**Figs. 11, 12**). BASIONYM: *Cynosurus aegyptius* L., Sp. Pl. 72. 1753.

Robust, weedy, mat-forming to erect grass. Culms rising to 55 cm tall, stoloniferous and rooting at the nodes, glabrous. Sheaths 1.4–5.5 cm long, glabrous or nearly so. Blades 1–21 cm long, 2–5(–6) mm wide, with long pustule-based hairs on both surfaces. Inflorescences 2–4 cm long, 2–5 cm wide, usually held well above the foliage, consisting of 3–4 stubby, digitately arranged spikes 0.9–3 cm long; rachises conspicuous, straight and extending past spikelets as sharp points. Spikelets ca. 3 mm long, in crowded, 2-rowed herringbone pattern on one side of rachis, few-flowered. Glumes broadly ovate, nearly equal, awned. Lemmas resembling the glumes, the larger ones awned. Palea small and hyaline, folded to match the lemma, with the keel of the palea nestled into the lemma. Caryopsis rugose.

Field recognition characters.—Spikes particularly short and stubby, with the rachis extending past the florets as a point; blades hirsute.

Common name.—Egyptian Grass.

Distribution and habitat.—*Dactyloctenium aegyptium* is an Old World weed encountered in abundance throughout Barbados on disturbed sites and on beaches.

Representative specimens: **BARBADOS. Christ Church:** Newton, roadside ditch, Aug 1940, *Gooding 505* (BAR); Silver Sands Park, beach, area in front of dunes, Oct 1996, *Forde 96-1* (BAR). **St. Andrew:** Along coastal highway between Barclays Park and Belleplaine, on base of rocky hills rising above beach, 4 Sep 1996, *Rogers 96-31* (BAR, FLAS). **St. George:** Gun Hill, dry ground, Jan 1939, *Allan 505* (BAR). **St. Joseph:** Tenby Bay House, "4.11.1901," *Freeman & Bovell 263* (BAR); Bathsheba, grass land by the sea, "7.1.18," *Royal Bot. Gard. Trinidad & Tobago 8683* (BAR). **St. Michael:** Government House, "20.9.01," *Bovell 263* (BAR). **St. Peter:** Beach, Speightstown, 29 Oct 1996, *Rogers 96-120* (BAR).

Dichanthium Willemet, Ann. Bot. (Usteri) 18:11. 1796.

Often-stoloniferous, often-sprawling grasses. Inflorescences racemose or paniculate clusters of fingerlike spikes. Spikelets in pairs—one sessile and fertile, the other pedicellate and sterile (in ours). Lowest pair of spikelets on the rachis awnless and sterile. Disarticulation between the 2 spikelets of the pair.

KEY TO *DICHANTHIUM* SPECIES IN BARBADOS

1. Culm with a densely pubescent region just below the inflorescence .... **Dichanthium aristatum**
1. Culm glabrous or nearly so immediately below the inflorescence .... **Dichanthium annulatum**

19. Dichanthium annulatum (Forssk.) Stapf, Fl. Trop. Africa 9:178. 1917. BASIONYM: *Andropogon annulatus* Forssk., Fl. Aegypt.-Arab. 173. 1775.

Culms to 0.75 m tall, tufted, with conspicuously bearded nodes having erect hairs. Sheaths 2–3 cm long. Blades 3–10 cm long, sometimes hirsute, especially so adaxially and basally. Inflorescences consisting of 1 or 2 spikes 1.5–5.5 cm long (excluding awns), the spikes devoid of spikelets near the base. Pedicellate spikelet represented by a pair of empty glumes 3.5–4 mm long, these rounded and with conspicuous nerves. Sessile spikelet 3 mm long, fertile, the large first glume the most conspicuous foliar organ in the spikelet, ciliate-pubescent, ca. 9-nerved,

FIG. 11. *Dactyloctenium aegyptium* (*Rogers 96-120*)
FIG. 12. *Dactyloctenium aegyptium* (*Forde 96-17*)

rounded apically. Second glume lanceolate, trinerved, hyaline and much smaller than the first glume. Fertile lemma reduced, bearing an awn to 17 mm long, the fertile palea not evident. Sterile floret represented by an inconspicuous hyaline lemma.

Field recognition characters.—Spikes single or paired, the glumes without pits (vs. these present in similar *Bothriochloa* species); nodes heavily bearded with erect hairs.

Distribution and habitat.—*Dichanthium annulatum* is an Old World pasture species well-established in pastures and in disturbed weedy sites, sometimes in lawns.

Representative specimens: **BARBADOS. Christ Church:** ABC Highway, roadside weed near Kingsland, glabrous below inflorescence, 11 Sep 1996, *Rogers 96-53* (BAR). **St. Andrew:** On dunes behind beach, between Belleplaine and Barclays Park, 18 Oct 1996, *Rogers 96-96* (BAR). **St. Joseph:** Canefield at top of Hackleton's Cliff, single raceme and pubescence at nodes reduced, 29 Nov 1996, *Rogers 96-140* (BAR). **St. Michael:** Codrington Expt. Station, roadsides, 21 Oct 1937, *McIntosh 442* (BAR).

20. Dichanthium aristatum (Poir.) C. E. Hubb., Bull. Misc. Inform. Kew 1939:654. 1939. **(Figs. 13, 14)**. BASIONYM: *Andropogon aristatum* Poir., Encycl., Suppl. 1:585. 1811.

Culms rising to 1 m tall, 0.5–1 mm in diameter at the base of the inflorescence, this region densely hirsute, bearded at the nodes, rooting at the lower nodes. Sheaths to 7.5 cm long with a tuft of long hairs at the junction with the blade, otherwise glabrous. Blades 5–18 cm long, up to 5 mm wide, scabrous marginally and sometimes on the abaxial midrib. Inflorescences made up of (1–)2–4 pubescent spikes 2–8 cm long. Sterile glumes of the pedicellate spikelet 3 mm long, conspicuous, elliptic, hirsute, obtuse or rounded to acute apically, sometimes apiculate, the longitudinal nerves prominent, the margin fringed. Fertile sessile spikelet with the 2 glumes resembling those of the pedicellate spikelet, the fertile lemma reduced, bearing an awn to 25 mm long, the fertile palea not evident. Sterile floret represented by a hyaline lemma.

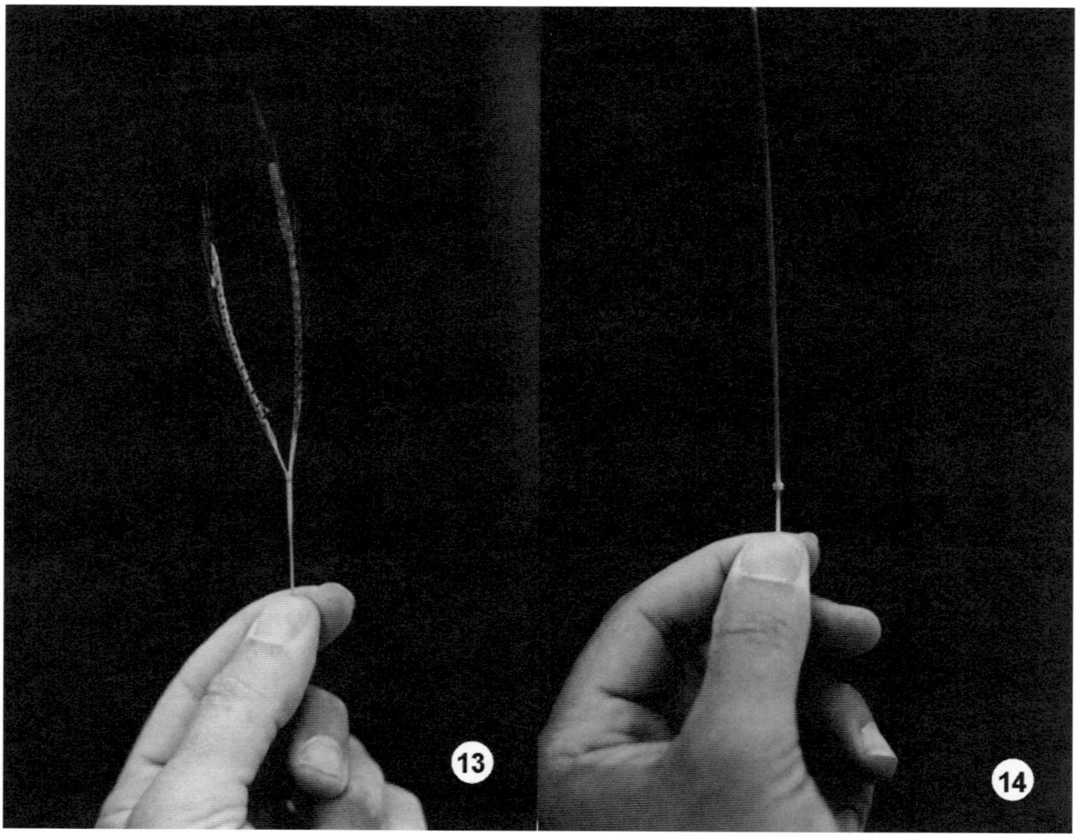

FIG. 13. *Dichanthium aristatum* (*Rogers 97-22*)
FIG. 14. *Dichanthium aristatum* (*Rogers 97-22*)

Field recognition characters.—Spikes 1–4, the area immediately beneath them on the inflorescence densely pubescent, the nodes bearded.

Distribution and habitat.—*Dichanthium aristatum* is an Old World pasture species abundant in pastures and in weedy places around Barbados.

Representative specimens: **BARBADOS. Christ Church:** Chancery Lane, meadow S of pond, 23 Nov 1996, *Rogers 96-135* (BAR, FLAS). **St. Michael:** Lower Estate, roadside gutter, Nov 1935, *Mayers 20* (BAR); Codrington Expt. Station, roadsides, poorer parts of sourgrass fields, 24 Oct 1937, *McIntosh 441/20a* (BAR); University of the West Indies campus, in neglected, dry lawn, 29 May 1997, *Rogers 97-22* (BAR).

Digitaria Haller, Hist. Stirp. Helv. 2:244. 1768, nom. cons.

Inflorescences umbellate or racemose (or paniculate with the spikes branched basally), having thin spicate branches, these generally with the rachis broadened. Spikelets in 2 rows, usually paired (in ours) with one sessile or short-pedicellate, the other long-pedicellate, each spikelet with 2 florets, the lower floret represented by a large lemma, its palea missing or very reduced. First glume very reduced (in ours) or missing. Second glume much shorter than the spikelet (in ours).

A large cosmopolitan genus.

Note that Gooding et al. (1965) casually mentioned *Digitaria horizontalis* Willd. as present in Barbados. There is no substantiation for this.

KEY TO *DIGITARIA* SPECIES IN BARBADOS

1. Inflorescence a plumose, feathery panicle .... **Digitaria insularis**
1. Inflorescence digitate 2
   2. Sterile lemma of the pedicellate spikelet having soft cilia mixed with stiff, pectinate bristles; sterile lemma of the sessile spikelet having the nerves evenly spaced .... **Digitaria bicornis**

2. Sterile lemma of the pedicellate spikelet having soft cilia but no stiff bristles; sterile lemma of the sessile spikelet having the nerves unevenly spaced .... **Digitaria ciliaris**

21. Digitaria bicornis (Lam.) Roem. & Schult., Syst. Veg. 2:470. 1817. BASIONYM: *Paspalum bicorne* Lam., Tabl. Encycl. 1:176. 1791.

Culms rising to ca. 30 cm tall, ca. 0.5 mm in diameter at the apex, tending to sprawl and root at the nodes, straw-colored when dry. Sheaths 1.5–7 cm long, hirsute to puberulent-glabrate, with cilia near the collar. Blades 2–4 cm long, to 3 mm wide, lanceolate or broadest at the base, the midvein prominent abaxially in the lower half of the blade, scabrous-margined, with highly variable pubescence, both surfaces hirsute, or puberulent, or glabrous. Inflorescences digitate with 2 or 3 spikes 3–8 cm long, the rachis broadened and serrulate. Spikelets 3–3.2 mm long, some subsessile, the others distinctly pedicellate, borne singly or paired, when paired, the pairs made up of one subsessile and one long-pedicelled spikelet. Subsessile spikelet with a minute scalelike first glume, the second glume ca. 3/4 the length of the spikelet, covered abaxially with long silky hairs, narrowly lanceolate; sterile lemma as long as the spikelet, narrowly lanceolate, covered with long silky hairs, these sometimes with more substantial bristles, 3- or 5-nerved, the nerves evenly spaced; fertile palea and lemma glabrous, lanceolate. Long-pedicelled spikelet with the glumes as on the subsessile spikelet, the sterile lemma covered with wide-spreading, long silky hairs mixed with longer, thicker, pectinate bristles; fertile palea and lemma as on subsessile spikelet.

Field recognition characters.—Digitate inflorescence with the spreading cilia and pectinate bristles on the spikelets visible to the naked eye.

Distribution and habitat.—U.S. to South America. *Digitaria bicornis* has not been reported previously for Barbados, where it is a weed of disturbed sites.

Comments.—Establishing the identities of the common upright, digitate *Digitaria* species in Barbados has been problematic, as Gould (1979) discussed, and as revealed by floristic literature on the Caribbean region. Older herbarium specimens from the region are often misidentified as *Digitaria sanguinalis* (L.) Scop., the Crab Grass familiar in cool climates. *Digitaria sanguinalis* is not authentically recorded from Barbados. A pair of poorly differentiated widespread weedy species is present in addition to the distinctive *D. insularis*. *Digitaria bicornis* and *D. ciliaris* are similar, both having digitate spikes bearing spikelet pairs with one sessile or nearly so and the other distinctly pedicellate. In both species the first glume is a small scale, and the second glume is silky pubescent and far shorter than the spikelet. Both have long pubescence on the margins of the sterile lemmas. The two remain confused in the literature, with different authors finding different and contradictory ways to distinguish them. Alleged characteristics by which *D. bicornis* differs from *D. ciliaris* include the former having evenly spaced veins on the sterile lemma of the sessile spikelet (vs. the veins crowded toward the margin in *D. ciliaris*), this lemma having pubescent margins (vs. glabrous margins), the first glume rounded to truncate and very short to nearly absent (vs. obtuse to acute and longer than 0.3 mm), the sterile lemma on the pedicellate spikelet with strongly spreading cilia (vs. cilia absent or more often failing to spread), and the sterile lemma of the pedicellate spikelet with long, thickened bristles among the cilia (vs. there being only soft, silky cilia).

Henrard (1950) examined these two species in depth. He found the critical pedicellate spikelets to be missing from the type of *D. bicornis*, and concluded that the type of *D. ciliaris* is lost. (A lectotype was designated by Veldkamp, 1973, fide Davidse & Pohl, 1994.) Henrard provided a detailed illustration of the *D. bicornis* type, as well as illustrations of *D. ciliaris* based on non-type material. Based on Henrard's discussion and illustrations coupled with my (GR) examination of extensive non-Barbados material at FLAS in addition to Barbados specimens, I find no differences in the first glumes. Consistent with Henrard (1950), I find *D. bicornis* to have the sterile lemma of the pedicellate spikelet bearing long, wide-spreading, soft cilia mixed with longer, thicker, stiffer bristles and to have the nerves on the sterile lemmas of the sessile spikelets evenly spaced. By contrast, *D. ciliaris* has the long soft cilia on the sterile lemma of the pedicellate spikelet more appressed to the margin, less conspicuous, and free of bristles. The nerves on the sterile lemma of the sessile spikelet are (with exceptions) uneven, with a wide space between the central nerve and the ones flanking it, the latter crowded toward the margin. (Henrard 1950, recognized a *D. bicornis* subsp. *gamblei* Henrard which lacks the marginal bristles.)

Representative specimens: **BARBADOS. Christ Church:** On side of street, across from 135 Atlantic Shores, 2 Aug 1997, *Rogers 97-30* (BAR, FLAS). **St. Andrew:** Belleplaine, grey dunes near bottom of slope, Apr 1940, *Gooding 446* (BAR).

22. Digitaria ciliaris (Retz.) Koeler, Descr. Gram. 27. 1802. BASIONYM: *Panicum ciliare* Retz., Observ. Bot. 4:16. 1786.

Digitaria adscendens (Kunth) Henrard, Blumea 1:92. 1934.

Culms upright or sprawling and rooting at the nodes. Sheaths to 15 cm long, hirsute to puberulent. Blades lanceolate (4–)6–17 cm long, 3–11 mm wide, scabrous-margined, hirsute to puberulent-glabrate on both surfaces. Inflorescences racemose or digitate, with (2–)5–15 spikes (5–)10–14.5 cm long, the rachis broadened and striate. Spikelets paired, one sessile or nearly so, the other pedicellate. Pedicellate spikelet with first glume 0.25–0.5 mm long, or this barely perceptible, having a truncate to acuminate apex; second glume far shorter than the spikelet, sometimes less than half its length, silky pubescent; sterile lemma strongly 5-nerved, as long as the spikelet, densely silky ciliate marginally; fertile lemma and palea hyaline and similar to each other. Sessile spikelet with first and second glumes as on pedicellate spikelet; sterile lemma elliptic-lanceolate, 5-nerved with the nerves usually unevenly spaced, the lateral nerves crowded toward the margin (or the nerves even), marginally ciliate; fertile lemma and palea similar to each other and hyaline.

Field recognition characters.—Leaf blades generally pubescent; inflorescences racemose-digitate with usually more than five spikes.

Distribution and habitat.—*Digitaria ciliaris* is an abundant and variable pantropical weed.

Representative specimens: **BARBADOS.** No locality data. "11.9.01," common weed, *Bovell 264* (BAR). **Christ Church:** Behind beach, sandy substrate, between Silver Sands Hotel and Silver Rock Hotel, 27 Sep 1996, *Rogers 96-70* (BAR). **St. Andrew:** Turner's Hall Woods, in opening, 17 Dec 1996, *Rogers 96-153* (BAR); Highway 2, just N of Chalky Mt. Road, near Haggats, 18 Sep 1996, *Rogers 96-55* (BAR). **St. George:** In broken aging pavement around oil pump, just N of Highway 5 intersection with road to Turnpike and Carmichael, 16 Oct 1996, *Rogers 96-84* (BAR). **St. John:** Wasteland, Codrington College, 21 Oct 1937, *McIntosh 446* (BAR). **St. Joseph:** Garden bed, Andrews Factory, Sep 1936, *Mayers 446* (BAR); Andrews, roadside, Nov 1935, *Mayers 446* (BAR) [note repeated species number]; on face of Hackleton's Cliff, in shade, 29 Nov 1996, *Rogers 96-144* (BAR); Rusher Gully, gully floor, Jun 1940, *Gooding 446* (BAR). **St. Peter:** Speightstown, on wet marly sand, 10 Oct 1996, *Rogers 96-76* (BAR, FLAS).

23. Digitaria insularis (L.) Fedde, Just's Bot. Jahresber. 31:778. 1904. BASIONYM: *Andropogon insulare* L., Syst. Nat. ed. 10, 2:1304. 1759. *Trichachne insularis* (L.) Nees, Fl. Bras. Enum. Pl. 2:86. 1829.

Culms to 1.5 m tall (but usually much shorter), ca.1–2 mm in diameter near the apex, striate when dry. Sheaths 4–13(–18) cm long, glabrous to abaxially hirsute, and ciliate around the collar. Blades 8–50 cm long, 3–16 mm wide, glabrous except for cilia at collar and sometimes on lower adaxial surface, usually tapered to the base, the midrib conspicuous on both surfaces, the margins serrulate. Inflorescences feathery paniculate plumes up to 30 cm long and 6 cm wide, with branches 14 cm long. Spikelets ca. 6 mm long, pedicellate, often but not always paired, when paired one on a much shorter pedicel than the other. First glume a tiny scale. Second glume ca. 3/4 the length of the spikelet, narrowly lanceolate, covered with long silky hairs. Sterile lemma resembling the second glume but longer. Fertile lemma much smaller, wrapped around the palea, glabrous, lanceolate.

Field recognition character.—Inflorescence a silky panicle.

Distribution and habitat.—*Digitaria insularis* ranges from the U.S. through the Caribbean to southern South America. It crops up in Barbados on highly diverse exposed sites.

Representative specimens: **BARBADOS. St. Andrew:** Turner's Hall, waste land near woods, May 1940, *Gooding 512* (BAR). **St. James:** Weed on disturbed ground at Port Vale Sugar Factory, 1 Sep 1997, *Rogers 97-33* (BAR). **St. Philip:** On rock wall along road immediately uphill from Crane Beach, 20 Oct 1996, *Rogers 96-88* (MICH); Three Houses, waste land near stream, Sep 1940, *Gooding 512* (BAR) [note repeated species number].

Dinebra Jacq., Fragm. Bot. 77. 1809.

A warm-climate Old World genus of three species.

24. **Dinebra retroflexa** (Vahl) Panz., Denkschr. Kaiserl. Akad. Wiss., Math.-Naturwiss. Kl. 4:270, t. 12, 1814. (**Figs. 15, 16**). Basionym: *Cynosurus retroflexus* Vahl, Symb. Bot. 2:20. 1791.

Culms to 81 cm long, sometimes purplish when dry, rooting at the lower nodes, sometimes rhizomatous. Sheaths 2.5–11 cm long, sometimes purplish when dry, laterally compressed, glabrous except for some cilia near the collar. Ligule to ca. 1.2–2 mm long, an irregular whitish membrane made up of laterally fused fibers. Blades 6–20(–28) cm long, 3–9 mm wide, usually widest just above the base (very narrowly lanceolate), mostly glabrous but with some adaxial cilia, and with some cilia near the collar. Inflorescence 9–18 cm long, up to 4 cm wide, consisting of a long central axis bearing wide-spreading, straight, unbranched side branches to 3.5 cm long, these diminishing in length from the base toward the apex of the inflorescence, the rachises broad and flattened. Spikelets ca. 2.6 mm long (excluding the long awns on the glumes), 3-flowered, each floret on a short stalk. Glumes 5–6 mm long, very narrow, subequal, subulate, 1–nerved, drawn out into an awn, scabrous on the keel. Lower and middle florets with similar lemmas, these ca. 2 mm long, 1–nerved, elliptic, delicate and associated with inconspicuous hyaline paleas; uppermost floret reduced.

Field recognition characters.—Sprawling weed with distinctive inflorescences: the straight side branches diminishing in length from the base to the apex of the inflorescence. Glumes exceptionally long and drawn out into awns.

Distribution and habitat.—This distinctive African weed has not been previously reported for Barbados nor from the Antilles. No reports of it in the New World are known to us. It occurs on seasonally moist clay.

Comments.—*Dinebra retroflexa* is readily recognized as an often prostrate grass with numerous, short, straight side branches on the main inflorescence axis. The awned glumes are far longer than the lemmas.

Representative specimens: **BARBADOS. Christ Church:** Weed in canefield along road connecting Newton with Lower Greys, sprawling weed on black clayish (manured and tilled) agricultural soil, 12 & 13 Nov 1997, *Rogers 97-53* (BAR, FLAS). **St. George:** Weed between canefield and road connecting highways 4 and 5 via Carmichael, black, montmorillonite-dominated agricultural soil, 6 Jan 1998, *Rogers 98-1* (BAR).

Echinochloa P. Beauv., Ess. Agrostogr. 53. 1812, nom. cons.

Weedy grasses tending toward thick culms. Inflorescences racemose or paniculate, with 2-flowered spikelets, the lower floret staminate or neuter, the upper one perfect. Lemma of upper floret indurate, smooth, with inrolled margins. Tips of lower and of upper lemmas apiculate.

A small weedy genus primarily of warm climates.

KEY TO *ECHINOCHLOA* SPECIES IN BARBADOS

1. Leaf blades under 6 mm wide, eligulate .... **Echinochloa colona**
1. Leaf blades broader than 1 cm, ligulate .... **Echinochloa polystachya**

25. **Echinochloa colona** (L.) Link, Hort. Berol. 2:209. 1833, as "*colonum*." Basionym: *Panicum colonum* L., Syst. Nat. ed. 10, 2:870. 1759.

Erect, weedy, nearly glabrous, variable grass. Culms up to 60 cm tall but often sprawling and rooting at the nodes. Sheaths to 14.5 cm long, sometimes ciliate at the collar, eligulate. Blades up to 20(–26) cm long and 5(–6) mm wide, thin-textured and flimsy upon drying, glabrous or abaxially with minute puberulence, smooth-margined or sometimes serrulate. Inflorescences 3.5–13 cm long and up to 1.5 cm wide, racemose, with 2–14 straight, ascending side branches 5–25(–29) cm long at intervals of 6–19(–30) mm, sometimes subopposite, densely covered with spikelets on one side, the rachis hispidulous. Spikelets 2.1–2.5 mm long. First glume half as long as the spikelet, broadly ovate with a slightly protruding mid-nerve, trinervate, hispidulous. Second glume as long as the spikelet, otherwise resembling the first glume. Lower (sterile) lemma elliptic-lanceolate, as long as the spikelet, its palea well-developed and as long as the spikelet, hyaline. Upper lemma elliptic-lanceolate, indurate.

FIG. 15. *Dinebra retroflexa* (*Rogers 97-53*)
FIG. 16. *Dinebra retroflexa* (*Rogers 97-53*)

Field recognition characters.—Blades thin, eligulate. Inflorescences racemose with the upswept branches densely and unilaterally covered with spikelets.

Distribution and habitat.—*Echinochloa colona* is a pervasive weed of moist sites in Barbados. Now worldwide in tropical and subtropical regions, it probably originated in the Old World (Pohl 1980).

Comments.—Authors differ on the use of "colonum" vs. "colona," apparently in part because the fundamental nature of the epithet is not clear (see a brief discussion in McVaugh 1983).

Representative specimens: **BARBADOS.** No locality data. *Herbarium, Dept. of Agriculture, Barbados No. 240* (BAR); ditches and damp trenches, Aug 1901, *Bovell 240* (BAR); Boarded Hall, weed, wet ditch, Oct 1939, *Allan 11a* (BAR). **St. Andrew:** "11.9.1904," *Murphy 240* (BAR); Highway 2 near Haggats, on stream bank, 4 Sep 1996, *Rogers 96-29* (BAR, FLAS). **St. Philip:** Three Houses, floor of stream, "water grass," May 1940, *Gooding 11* (BAR); Crane, summit of cliff, undated, *Freeman 240* (BAR). **St. Thomas:** Roadside gutter, Strong Hope, Oct 1935, *Mayers 11* (BAR).

26. Echinochloa polystachya (Kunth) Hitchc., Contr. U.S. Natl. Herb. 22:135. 1920. BASIONYM: *Oplismenus polystachyus* Kunth, Nov. Gen. Sp. 1:107. 1816.

Culms upright. Sheaths glabrous. Ligule a band of cilia ca. 3 mm long. Blades to 40 cm long and 17 mm wide, narrowly elliptic-lanceolate, with a conspicuous white midrib (dry) and ca. 10–12 strongly developed longitudinal nerves, scabrous-margined, glabrous except for a small patch of pubescence immediately distal to the ligule. Inflorescences to 28 cm long, 4–7 cm wide, pyramidal, the straight branches to ca. 8.5 cm long, ascending, flanged, the flanges hispid, the branches densely covered with spikelets ca. 5 mm long (excluding glumes). First glume a little over half the length of the spikelet, ovate, wrapped basally around the spikelet (excluding the awns). Second glume as long as the spikelet, short-awned, 5(–7)-veined, hispid on the veins and puberulent between the veins. Lower lemma lanceolate, bearing an apical awn 5–6 mm long, resembling the glumes in texture, hispid on the veins. Upper lemma whitish and smooth, shorter than the lower one.

Field recognition characters.—Large coarse grass having conspicuous white midribs in the leaf blades; inflorescences large (generally over 20 cm long), pyramidal with straight ascending branches; spikelets awned.

Distribution and habitat.—Although distributed from the U.S. to southern South America, *Echinochloa polystachya* has not been previously reported for Barbados. At a glance it resembles *Sorghum halepense* in several characteristics but differs by having the spikelets awned on the second glume and awned on a well-developed lemma (as opposed to being awned solely on a much-reduced distinctive bilobed lemma), and by having distinctive stout hairs on the veins of the glumes and the veins of the lower lemma.

Representative specimens: **BARBADOS. St. Philip:** Pasture at Golden Grove, 6 Dec 1996, *Rogers 96-145* (BAR, FLAS).

Eleusine Gaertn., Fruct. Sem. Pl. 1:7. 1788.

Spikelets sessile in 2 rows on digitate branches (often with one branch on the culm below the others). Spikelets with 3 or more florets, disarticulating above the glumes and between the florets.

A small primarily Old World genus.

27. **Eleusine indica** (L.) Gaertn., Fruct. Sem. Pl. 1:8. 1788. **(Fig. 10).** BASIONYM: *Cynosurus indicus* L., Sp. Pl. 72–73. 1753.

Large, coarse, tufted, weedy grasses. Culms to ca. 70 cm tall. Blades to 35 cm long and 2–7 mm wide, smooth-margined, folded upon drying, nearly glabrous except for long silky adaxial (and sometimes abaxial) hairs especially near the basal margins, the sheaths flattened, equitant, sometimes broader than the leaf blades, to 10.5 cm long, long-hirsute (especially toward the apex and at the collar) to glabrous. Inflorescence digitate with 2–6 straight spikes 1.5–12.5 cm long and 2–6 mm wide, often with one spike borne alone below the others. Spikelets 3–6 mm long, multiflowered, in a chevron pattern along one side of a flattened rachis. Glumes elliptic-lanceolate, shorter than the spikelets, the first glume a little smaller than the second glume; lemmas boat-shaped (ovate when unfolded), paleas pleated, almost as large as the lemmas, puberulent on the pleats.

Field recognition characters.—Large weedy grass, with more or less digitate inflorescences having unusually large (3–6 mm long) spikelets arranged in a chevron pattern; frequently with one spike jutting out alone from the culm beneath the otherwise digitate cluster.

Common names.—Man Grass, Dutch Grass.

Distribution and habitat.—This coarse aggressive Old World weed is abundant throughout Barbados on disturbed sites and on beaches.

Representative specimens: **BARBADOS.** No locality data. Weed, "3,7,1901," [also Aug 1902] *Bovell 233* (BAR); Feb 1937, *Allan 19b* (BAR). **Christ Church:** Newton, waste ground to W of factory, June 1940, *Gooding 19* (BAR); Silver Sands Park, 19 Jan 1997, *Rogers 97-7* (BAR); Graeme Hall Swamp, open grassland, 25 Feb 1980, *Riven-Ramsey 46* (BAR); Silversands Beach Park, open area in front of dune, Oct 1996, *Forde 96-3* (BAR). **St. Andrew:** Highway 2 near Haggats, stream bank, 4 Sep 1996, *Rogers 96-30* (BAR). **St. Joseph:** Andrews Plantation., Oct 1935, *Mayers 19* (BAR); Andrews Plantation, cart-road, Jul 1936, *Mayers 19a* (BAR); weed at Andromeda Botanic Gardens, 27 Jan 1998, *Rogers 98-6* (BAR).

Eragrostis Wolf, Gen. Pl. 23. 1776.

Inflorescences paniculate. Spikelets usually multiflowered, disarticulating above the glumes. Lemmas trinerved. Paleas well-developed, plicate, often scabrous to ciliate on the keels.

A weedy, cosmopolitan genus of some 350 species.

In addition to the species treated here, Gooding et al. listed *Eragrostis glomerata* (Walter) L.H. Dewey, Contr. U.S. Natl. Herb 2:543. 1894 as having once been found in Barbados in the "swamp at the Mental Hospital." There is no specimen citation, there is no specimen in the

FIG. 10. *Eleusine indica* (Rogers 98-6)

Barbados National Herbarium, we have not encountered this species, and have we have not seen material of it. The "Mental Hospital" is still present in a residential area with no swamp.

KEY TO *ERAGROSTIS* SPECIES IN BARBADOS

1. Paleas with the keels ciliate (the cilia longer than the width of the folded palea) 2
1. Paleas with the keels glabrous to merely puberulent-scabrous (if trichomes present, these far shorter than the folded width of the palea) 3
   2. Inflorescence an open, airy panicle .... **Eragrostis amabilis**
   2. Inflorescence a tightly congested, spicate panicle .... **Eragrostis ciliaris**
      3. "Panicle elongate, contracted, the minute spikelets crowded or glomerate" (fide Gooding et al. 1965) .... **Eragrostis glomerata** (see comments above)
      3. Panicle open, expanded 4
         4. Spikelets usually over 7 mm long with usually over 8 florets .... **Eragrostis acutiflora**
         4. Spikelets shorter than 7 mm with usually fewer than 7 florets 5
            5. Spikelets ca. 1 mm wide .... **Eragrostis pilosa**
            5. Spikelets 1.5–2 mm wide .... **Eragrostis pectinacea var. miserrima**

28. Eragrostis acutiflora (Kunth) Nees, Bras. Enum. Pl. 501. 1829. BASIONYM: *Poa acutiflora* Kunth, Nov. Gen. Sp. 1:161. 1816.

Tufted grass 25–80 cm tall. Sheaths 4–11(–14) cm long, silky pubescent marginally and sometimes abaxially. Blades 2.5–4.2 cm long, 1–2 mm wide, acicular, folded-involute upon drying, adaxially scabrous and sometimes with silky hairs toward the adaxial base, bearded at the collar. Inflorescences open, 7–30 cm long, 3–11 cm wide, paniculate, with white-ciliate beards at the bases of the branches. Spikelets (5–)7–14 mm long, with up to (5–)8–12 florets. Glumes 0.5–1 mm long, delicate and caducous, the second glume longer than the first. Lemmas (1.5–)2–2.3 mm long, lanceolate, 1–nerved, scabrous on the keel. Palea pleated, shorter than the lemma, scabrous on the keels.

Field recognition characters.—Inflorescences open-paniculate having spikelets often exceeding 0.7 cm long and usually bearing over 8 florets.

Distribution and habitat.—*Eragrostis acutiflora* is distributed in Central and South America, with a small number of records in the West Indies, including Trinidad. It has not been previously reported for Barbados. Despite this, the species is abundant throughout the island, turning up in disturbed uncultivated sites as well as in lawns and athletic fields.

Comments.—This species differs from the abundant and similar *E. amabilis* by having spikelets mostly over 7 mm long (vs. mostly under 2 mm long) and with the paleas merely scabrous (vs. ciliate) on the keels. Likewise similar, *E. pilosa* has shorter spikelets (mostly 2–5 mm vs. usually longer than 7 mm) and fewer than 7 (vs. usually over 8) florets per spikelet.

Representative specimens: **BARBADOS. Christ Church:** Chancery Lane, meadow between pond and beach, low, wet, marly ground, 23 Nov 1996, *Rogers 96-133* (BAR, FLAS, MICH). **St. Joseph:** Sunny meadow on hill above Bathsheba, 18 Sep 1997, *Rogers 97-43* (BAR); Highway 3, roadside weed, between Horse Hill and cutoff to Castle Grant (on the roadside in front of Grantley Adams Memorial School), 11 Sep 1997, *Rogers 97-41* (BAR).

29. Eragrostis amabilis (L.) Wight & Arn. ex Nees, Bot. Beechey Voy. 251. 1838. Basionym: *Poa amabilis* L., Sp. Pl. 68. 1753.

Eragrostis tenella (L.) P. Beauv. ex Roem. & Schult., Syst. Veg. 2:576. 1817. *Poa tenella* L., Sp. Pl. 69. 1753.

Culms to 31 cm tall. Sheaths 0.8–2.5 cm long, glabrous. Blades 2.5–6 cm long, ca. 1(–2) mm wide, acicular, glabrous except for a beard at the collar. Inflorescences 2–10 cm long, 1–5 cm wide, open, airy, pyramidal panicles, the axes thin, wiry. Spikelets 1–2 mm long with 2–10 florets. Glumes ovate to lanceolate, similar to the lemmas but only 1–nerved, the first glume shorter than the second glume. Lemma 0.5–1 mm long, boat-shaped, scabrous-puberulent on the keel, trinerved. Palea pleated, the keels with stiff, erect cilia ca. 0.5 mm long.

Field recognition characters.—Acicular narrow blades; open, airy panicle; spikelets under 2 mm long, with the paleas bearing large, erect, conspicuous cilia.

Distribution and habitat.—An abundant weed in Barbados, this species is an Old World introduction.

Comments.—This grass is generally known as "*Eragrostis tenella*." The name change comes from Webster & Peterson (1996). It differs from the similar *E. pilosa* by having spikelets under 2 mm long vs. usually over 2 mm long and ciliate (vs. scabrous-puberulent) paleas. *Sporobolus tenuissimus* is superficially similar but differs in numerous characters, including having just one floret per spikelet.

Representative specimens: **BARBADOS. St. George:** In broken aging pavement around oil pump; just N of Highway 5 intersection with road to Turnpike & Carmichael, 16 Oct 1996, *Rogers 96-86* (BAR). **St. Joseph:** Andromeda Botanic Gardens, weed growing on cinder blocks, 22 Dec 1997, *Rogers 97-69* (BAR). **St. Michael:** Weed on the University of the West Indies Campus, full sun, disturbed ground, 20 Aug 1996, *Rogers 96-12* (BAR); Codrington House, 1935, *Mayers 316* (BAR); Stafford House grounds, wayside moist places, Jan 1937, *Allan 316a* (BAR).

30. Eragrostis ciliaris (L.) R. Br., Narr. Exped. Zaire 478. 1818. Basionym: *Poa ciliaris* L., Syst. Nat. ed. 10, 2:875. 1759.

Culms 14–34 cm tall, upright, tufted. Sheaths to ca. 4.5 cm long, with long cilia toward the apex. Blades 2.5–11 cm long, 0.5–5 mm wide, glabrous except for long cilia at the collar. Inflorescences 2.7–12 cm long, 4–7(–14) mm wide, usually tightly congested (rarely with main axes spreading at acute angles), narrow, spikelike, often interrupted, paniculate. Spikelets 1–2.5 mm long with 4–7(–9) florets. Glumes lanceolate, uninerved with the nerve protruding as a tiny mucro, the first glume shorter than the second glume, this scabrous on the keel. Lemmas lance-elliptic, larger than the glumes, abaxially ciliate. Paleas pleated with numerous, conspicuous cilia up to 0.75 mm long on the pleats.

Field recognition characters.—Blades usually acicular or nearly so; inflorescences reminiscent in shape and size of pencils; paleas with long cilia.

Distribution and habitat.—This probable Old World species is widely naturalized and is abundant in Barbados on disturbed weedy sites.

Comments.—Differing from *E. tenella* by having a tightly congested inflorescence generally under 7 mm in diameter (vs. an open inflorescence over 1 cm in diameter).

Representative specimens: **BARBADOS.** Parish not specified. Clydesdale, 24 Aug 1901, *Freeman 270* (BAR). **Christ Church:** Maxwell's Coast, roadside near beach, May 1940, *Gooding 16* (BAR). **St. George:** In broken aging pavement around oil pump, just N of Highway 5 intersection with road to Turnpike and Carmichael, 16 Oct 1996, *Rogers 96-84, 96-85* (BAR, FLAS). **St. Joseph:** Andrews Plantation, rocky waste, Nov 1935, *Mayers 16* (BAR). **St. Michael:** Codrington, 24 Oct 1905, *Herbarium Dept. of Agriculture, Barbados 270* (BAR).

31. **Eragrostis pectinacea** (Michx.) Nees, Fl. Afr. Austral. Ill. 406. 1841 var. **miserrima** (E. Fourn.) Reeder, Phytologia 60:154. 1986. BASIONYM: *Eragrostis purshii* Schrad. var. *miserrima* E. Fourn., Mexic. Pl. 2:116. 1886.

Eragrostis tephrosanthos Schult., Mant. 2:316. 1824.

Culms upright, tufted, to 17 cm tall. Sheaths 1.5–3 cm long. Blades linear, 3–6 cm long, to ca. 1 mm wide. Inflorescences to 10 cm long and 1.5–3 cm wide, open-paniculate. Spikelets 3–5 mm long, 1.5–2 mm wide, bearing 2–5(–6) florets. Glumes elliptic-lanceolate, uninerved, the first glume shorter than the second glume, and this shorter than the lemmas. Lemmas lanceolate, trinerved. Paleas puberulent on the folds (the hairs much shorter than the folded width of the palea).

Field recognition characters.—Blades very narrow (ca. 1 mm wide); spikelets wider than 1.5 mm.

Distribution and habitat.—Ranging from the U.S. to Brazil, this species is known from Barbados from one collection almost a century old.

Comments.—Our description comes from the only Barbados specimen. *Eragrostis pectinacea* var. *miserrima* differs from the very similar *E. pilosa* by having spikelets 1.5–2 mm vs. ca. 1 mm wide.

Representative specimen: **BARBADOS. St. Philip:** Dodds, 13 Sep 1901, *Bovell 343* (BAR).

32. **Eragrostis pilosa** (L.) P. Beauv., Ess. Agrostogr. 71, 162, 175. 1812. BASIONYM: *Poa pilosa* L., Sp. Pl. 68. 1753.

Culm to ca. 40 cm tall with the nodes darkened. Blades acicular, 2–15 cm long, 0.5–3 mm wide, glabrous except for a beard of long silky cilia at the collar. Sheaths 1–10 cm long, glabrous. Inflorescences to 19 cm long and 8.5 cm wide, open, loose, wiry, paniculate. Spikelets 2–5(–7) mm long, ca. 1 mm wide on pedicels generally longer than the spikelets, bearing (2–)4–7 florets ca. 1.5 mm long. Glumes lanceolate, uninerved, the first glume much shorter than the second glume, and this much shorter than the florets. Lemmas lanceolate, trinerved, the larger ones scabrous on the keels. Paleas scabrous-puberulent on the keels.

Field recognition characters.—Leaves acicular; panicles open, airy, and wiry; paleas puberulent-scabrous (vs. ciliate in the similar *E. amabilis*). Spikelets ca. 1 mm wide, shorter than their pedicels and spreading (vs. 1.5–2 mm wide, longer than their pedicels and appressed in the similar *E. pectinacea* var. *miserrima*).

Distribution and habitat.—*Eragrostis pilosa* is a European species now in the U.S., Mexico, Caribbean, and Central America.

Representative specimens: **BARBADOS.** No locality data. 1902, *Bovell 495* (BAR). **Christ Church:** Newton, wayside ditch, Jul 1940, *Gooding 13* (BAR); sunny, exposed, highly disturbed waste ground in the Newton Factory complex, 12 Nov 1997, *Rogers 97-51* (BAR, FLAS). **St. George:** [Gun Hill], Rose Cottage, dry situation, weed, Aug 1938, *Allan 13* (BAR). **St. Joseph:** Andrews Plantation, Nov 1935, *Mayers 13* (BAR). **St. Philip:** Hampton, "11.9.01," *Bovell 495* (BAR).

Eriochloa Kunth in H.B.K., Nov. Gen. Sp. 1:94. 1816.

Inflorescences paniculate. Spikelets with a beadlike swelling at their base, having 2 florets, the upper one perfect, the lower one staminate or neuter. First glume imperceptible or membranous and fused with and surrounding the rachis. Second glume and sterile lemma of similar length. Lemma of fertile floret mucronate or with a short awn.

A small genus primarily from warm climates in the New World.

33. **Eriochloa punctata** (L.) Desv., Prodr. Pl. Ind. Occid. 5. 1825. Basionym: *Milium punctatum* L., Syst. Nat. ed 10, 2:872. 1759.

Culms up to at least 50 cm tall, having puberulence overtopped by a layer of much longer trichomes, the nodes puberulent. Sheaths 6–8 cm long, flattened, puberulent becoming glabrous. Blades to 12 cm long and 5 mm wide, glabrous except for scabrous margins. Inflorescence 5 cm long and 1.3 cm wide, with ca. 4 spicate side branches 4 cm long, the rachises heavily puberulent. Spikelets ca. 4.5 mm long, strigose. First glume a collarlike swelling around the rachis beneath the spikelet. Second glume lanceolate with a long-attenuate apex, strigose, 5-nerved, wrapped around the rest of the spikelet. Sterile lemma resembling the second glume though a little smaller. Fertile lemma white, indurate, rounded, tipped with an awn ca. 1 mm long.

Field recognition characters.—Bilayered pubescence on culms (with puberulent nodes); panicles with unbranched side branches, pubescent rachises and spikelets, and awned indurate lemmas.

Distribution and habitat.—*Eriochloa punctata* occurs from the U.S. to Argentina, and in Barbados is known only from Turner's Hall Woods.

Representative specimens: **BARBADOS. St. Andrew:** Turner Hall, 26 Sep 1906, *Dash 602* (BAR); Turner's Hall Woods, 4 June 1998, *Rogers 98-39* (BAR-sterile).

Eustachys Desv., Nouv. Bull. Sci. Soc. Philom. Paris 2:188. 1810.

Culms hollow. Spikes digitate (in ours). Spikelets borne perpendicular to and beneath the triangular rachis, laterally compressed with keeled glumes and lemmas, the second glume emarginate, short-awned, with 2 florets, the lower one fertile, the upper one represented by a much-reduced, truncate sterile lemma. Fertile lemma awnless, shiny and dark brown.

Eustachys is a small New World genus closely related to and sometimes merged with *Chloris*. *Eustachys* differs from *Chloris* by having florets essentially without awns (the second glume has a very short awn).

34. **Eustachys petraea** (Sw.) Desv., Nouv. Bull. Sci. Soc. Philom. Paris 2:189. 1810. (**Figs. 17–19**). Basionym: *Chloris petraea* Sw., Prodr. 25. 1788.

Culms rising (rarely) to 72 cm tall, 1–1.2 mm in diameter apically, stiff, upright. Sheaths up to 9 cm long, often broader than the blades, sharply keeled, tightly imbricate at the base of the plant, glabrous. Blades 8–24 cm long, 2–5 mm wide, ligulate with rounded or acute apices, glabrous except for the scabrous to hispidulous apical margins. Inflorescences held well above foliage, digitate with 3–5 spikes 2.5–8 cm long. Rachis triangular and grooved, hispidulous, bearing the spikelets in 2 rows on the lower surface. Spikelets 2.2–2.5 mm long, inserted at right angles to the rachis. First glume lance-deltoid, scabrous on the midrib. Second glume elliptic-oblanceolate, scabrous on the midrib, emarginate with a tiny awn (ca. 0.3 mm long) in the cleft.

Field recognition characters.—Rigidly upright beach species with tightly clustered equitant sheaths; spikes digitate, bearing patent spikelets. Awnless except for inconspicuous tooth on second glume.

Distribution and habitat.—*Eustachys petraea* is a rare species in Barbados, known only from a short stretch of beach in St. Lawrence Gap on the South Coast. It is found from the southern U.S. to Argentina.

Representative specimens: **BARBADOS. Christ Church:** Maxwell's Coast, beach near "Nova Lisboa," May 1940, *Gooding 313* (BAR); Maxwell Coast, seashore, weedy grass, Jan 1937, *Allan 313* (BAR); St. Lawrence Gap, Casuarina Beach, on beach sand at top of beach, 27 Aug 1996, *Rogers 96-25* (BAR); St. Lawrence Gap, 14 Nov 1996, *Forde 96-14* (BAR, FLAS).

Gynerium Willd. ex P. Beauv., Ess. Agrostogr. 138. 1812.

A monotypic neotropical genus.

35. **Gynerium sagittatum** (Aubl.) P. Beauv., Ess. Agrostogr. 138. 1812. (**Fig. 20**). Basionym: *Saccharum sagittatum* Aubl., Hist. Pl. Guiane 1:50. 1775.

Enormous dioecious grasses. Culms 7 m tall. Sheaths distichous, fringed with cilia, narrow-

FIG. 17. *Eustachys petraea* (*Forde 96-14*)
FIG. 18. *Eustachys petraea* (*Forde 96-14*)

ing toward the apex. Blades ca. 130 cm long and 6 cm wide, sharply serrulate (to strigose) marginally, long-attenuate, borne in a flat fan-shaped arrangement with the blade insertions separated evenly by ca. 3 cm, adaxially sometimes hirsute. Panicles to 1.5 m long, multibranched, feathery, upswept. Staminate spikelets (1–)2(–3)-flowered, 3.3–4 mm long, with membranous, lanceolate, uninerved glumes shorter than the spikelet, the first glume a little longer than the

FIG. 19. *Eustachys petraea* (*Forde 96-14*)
FIG. 20. *Gynerium sagittatum* (*Rogers 97-20*)

second glume; one floret below the other, the lower one staminate or with an empty lemma and palea, the upper one similar and consistently staminate. Lemmas and paleas lanceolate, the lemma with a prominent midnerve and a long attenuate apex. Pistillate spikelets not seen.

Field recognition characters.—Grass several meters tall resembling Sugar Cane and having leaf blades in a fanlike pattern; inflorescences over 1 m long.

Common name.—Wild Cane.

Distribution and habitat.—*Gynerium* forms thickets in water courses in the northern parishes of Barbados. Its distribution extends from Mexico to Paraguay.

Comments.—All Barbados specimens are staminate, suggesting clonality.

Representative specimens: **BARBADOS. St. Andrew:** Along road from Sedge Pond at first stream crossing just W of Bawdens, very common in this vicinity along streams, 5 May 1997, *Rogers 97-20* (BAR); streambank 0.6 km along paved road inland from Haggats Agricultural Station, near the traffic roundabout, staminate, long post-flowering, 9 Apr 1998, *Rogers 98-26* (BAR); on streambank ca. 1 km upstream from Walkers Bridge, staminate, 20 Apr 1998, *Rogers 98-27* (BAR) and *Rogers 98-28* (BAR, staminate). **St. Joseph:** In stream bed where crossed by East Coast Road 2 km S of Bathsheba, 7 m tall, the inflorescence to 1.5 m tall, staminate, 1 Sep 1997, *Rogers 97-34* (BAR). **St. Philip:** Dodds, common in the Scotland District [true, though Dodds, St. Philip, is not in the Scotland District], "5.8.01," *Bovell 250* (BAR).

Ichnanthus P. Beauv., Ess. Agrostogr. 56. 1812.

Broad-leaved grasses with a tendency to sprawl. Sheaths shorter then the internodes. Inflorescences paniculate. Spikelets paired and 2-flowered, the upper floret pedicellate and perfect, the lower one staminate or sterile. Similar to *Panicum* but with the upper lemma basally winged.

A small primarily warm-climate genus.

36. **Ichnanthus pallens** (Sw.) Munro ex Benth., Fl. Hongk. 414. 1861. (**Figs. 21, 22**). BASIONYM: *Panicum pallens* Sw., Prodr. 23. 1788.

Creeping shade-loving grass. Culms rising to 30 cm tall, ca. 1 mm (or a little less) in diameter apically, often with roots at the nodes. Sheaths 0.5–4.5 cm long, shorter than the blades and shorter then the internodes, with a marginal fringe of pubescence. Blades 3.5–11.5 cm long, 0.8–2.7 cm wide, lanceolate, with about 8 or 10 arcuate major lateral nerves, numerous minor lateral nerves, and a conspicuous midnerve, acuminate apically, the blade base auriculate, puberulent at the collar. Inflorescences to 5–12 cm long, 2.8–6 cm wide, with straight ascending primary branches and with far shorter secondary branches, sometimes with reduced axillary inflorescences. Spikelets 3.5–4 mm long, in unequally pedicellate pairs. First glume uninerved, lanceolate, shorter than the second glume. Second glume lanceolate, 5-nerved, as long as the spikelet. Lower lemma sterile, lanceolate, very similar to the second glume. Lower palea elliptic, hyaline. Upper lemma half as long as the lower one, bony and glossy near-white, elliptic, rounded abaxially, wrapped around its indurate palea, winged at the base.

Field recognition characters.—Creeping, broad-leaved, shade-loving grass with auriculate blade bases; sheaths short; panicles pyramidal. Upper florets indurate and shiny white, with the upper lemma basally winged.

Distribution and habitat.—*Ichnanthus pallens* is a species of damp forest sites from Central America to Argentina. In Barbados it occurs in Turner's Hall Woods, Welchman Hall Gully, and no doubt additional gullies.

Representative specimens: **BARBADOS.** No collection data. *Herbarium, Dept. Of Agriculture, Barbados s.n.* (BAR) & *458* (BAR). **St. Andrew:** Turner's Hall Woods, along dirt road in shade, 27 Jan 1998, *Rogers 98-13* (BAR, FLAS). **St. Joseph:** Hill above Bathsheba, shade under trees, Jan 1937, *Allan 320* (BAR). **St. Thomas:** Welchman Hall Gully, 24 Feb 1999, *Rogers 99-29* (BAR).

Lasiacis (Griseb.) Hitchc., Contr. U.S. Natl. Herb. 15:16. 1910.

Grasses with broad leaf blades, often sprawling or vinelike in habit. Inflorescences paniculate. Spikelets broad and inserted obliquely, 2-flowered, awnless. Lower floret neuter, the upper one indurate, its palea with a tuft of hairs.

A small genus restricted to warm regions in the New World.

KEY TO *LASIACIS* SPECIES IN BARBADOS

1. Blades narrower than 1 cm; panicle shorter than 5 cm .... **Lasiacis divaricata**
1. Blades broader than 1 cm; panicle longer than 5 cm .... **Lasiacis sorghoidea**

37. Lasiacis divaricata (L.) Hitchc., Contr. U. S. Natl. Herb. 15:16. 1910. BASIONYM: *Panicum divaricatum* L., Syst. Nat. ed. 10, 2:871. 1759.

Sprawling, shade-loving grass. Culms woody and glossy, having swollen nodes. Sheaths 15–20 mm long. Blades 5–10 cm long, 2.5–10 mm wide, narrowly lanceolate or linear, scabrous-margined, with a conspicuous midvein and 2–6 lateral nerves. Inflorescences 4–5 cm long, 2.5–3 cm wide, open panicles, often with one foliage leaf inserted above the lowest main branch, these inserted at right angles, bearing 1–3(–4) spikelets each. Spikelets 3–4.5 mm long, ca. 2 mm wide, inserted obliquely. First glume broadly ovate, multinerved, about half as long as the spikelet, with an apical tuft of hairs. Second glume elliptic with truncate or obtuse apex, as long as the spikelet, multinerved. Lower lemma closely resembling the second glume, empty but associated with a membranous palea. Upper lemma and palea indurate and glossy. Mature spikelets black.

Field recognition characters.—Sprawling shade-loving grass with leaf blades narrower than 1 cm and panicles under 5 cm long; mature spikelets black.

Distribution and habitat.—*Lasiacis divaricata* is indigenous to the greater Caribbean Basin. It is smaller and apparently less abundant locally than *L. sorghoidea.*

Representative specimens: **BARBADOS.** No collection data. *Herbarium, Dept. Of Agriculture, Barbados 414* (BAR). Parish not specified. Russia Gully, shade of trees, Nov 1938, *Allan 3a* (BAR) ["Rusher" Gully is in St. Joseph Parish]. **St. Andrew:** Turner's Hall Woods, deep shade, 17 Dec 1996, *Rogers 96-157* (BAR). **St. James:** Plum Tree Gully, gully margin in shade, 2 Dec 1989, *Carrington 1110* (BAR). **St. John:** Newcastle Wood, "5.11.06," *Dash 414* (BAR); floor of wood in Bath Gully, Nov 1935, *McIntosh 3* (BAR) [that this collection was made in St. John Parish is probable].

38. Lasiacis sorghoidea (Desv.) Hitchc. & Chase, Contr. U.S. Natl. Herb. 18:338. 1917. **(Figs. 23, 24).** BASIONYM: *Panicum sorghoideum* Desv., Prodr. Pl. Ind. Occid. 10. 1825.

FIG. 21. *Ichnanthus pallens* (*Rogers 98-13*)
FIG. 22. *Ichnanthus pallens* (*Rogers 98-13*)

Shade-loving scrambling or climbing grasses. Culms as tall as 2 m. Sheaths 3–5 cm long. Blades 7.5–15.5 cm long, 11–25 mm wide, a little asymmetric, elliptic or lanceolate, with a conspicuous midnerve and 9–12 lateral nerves, the apex acuminate and the base gradually narrowed to an uneven pseudopetiolate insertion, imbricated, hirsute. Inflorescences 6–16 cm long, 1.5–11 cm wide, open-paniculate. Spikelets 3–4 mm long, 2–4 mm wide, uncrowded, inserted obliquely on long, wiry pedicels. First glume broadly ovate-deltoid, half the length of the spikelet, tipped with a clump of white hairs, scabrous abaxially on the major nerves. Second glume as long as and partly wrapped around the spikelet, multiveined, elliptic with a truncate apex tipped with a white beard. Lower lemma empty but with a small, hyaline palea, resembling the second glume. Upper lemma resembling the lower one and the second glume but shorter, the upper lemma and its palea tipped with white hairs.

Field recognition characters.—Sprawling to climbing shade-loving grass, the first glume tipped with white hairs; mature spikelets black.

Distribution and habitat.—*Lasiacis sorghoidea* ranges from Mexico to South America, including the West Indies.

Representative specimens: **BARBADOS.** Parish not specified. Russia Gully, Nov 1938, *Allan 508* (BAR) ["Rusher" Gully is in St. Joseph Parish]. **St. Andrew:** Turner's Hall Woods, "14.12.02," *Bovell & Freeman 414* (BAR); Turner's Hall Woods, deep shade, 17 Dec 1996, *Rogers 96-154* (BAR, FLAS). **St. Joseph:** Hill above Bathsheba, shady, climbing in small trees, Jan 1937, *Allan 319* (BAR). **St. Peter:** Top of Cherry Tree Hill, deep shade on forest floor, rising to 2 m tall, some climbing trees, 17 Dec 1997, *Rogers 97-65* (BAR).

Leptochloa P. Beauv., Ess. Agrostogr. 71. 1812.

Inflorescences with slender, wandlike primary branches inserted along a vertical main axis. Spikelets multiflowered, the glumes thin with one or 3 nerves.

A small genus found around the globe in warm climates.

FIG. 23. *Lasiacis sorghoidea* (*Rogers 96-154*)
FIG. 24. *Lasiacis sorghoidea* (*Rogers 96-154*)

KEY TO *LEPTOCHLOA* SPECIES IN BARBADOS

1. Lemmas shorter than 1.7 mm; sheaths hirsute; spikelets usually with 1–3 florets .... **Leptochloa mucronata**
1. Lemmas longer than 1.7 mm; sheaths glabrous; spikelets usually with more than 3 florets ..... **Leptochloa virgata**

39. Leptochloa mucronata (Michx.) Kunth, Revis Gramin. 1:91. 1829. BASIONYM: *Eleusine mucronata* Michx., Fl. Bor.-Amer. 1:65. 1803.

Leptochloa filiformis (Pers.) P. Beauv., Ess. Agrostogr. 71, 163, 166. 1812.

Culms to ca. 1 m tall, upright to somewhat sprawling, with darkened internodes, sometimes rooting at the lower nodes, becoming glossy and a little woody. Sheaths 2.8–10 cm long, usually hirsute with pustule-based hairs. Blades 5–18(–24) cm long and 3–11 mm wide, flat, usually widest at or a little below the middle, glabrous or nearly so (occasionally at the bases briefly continuing the pubescence pattern of the sheaths), usually scabrous-margined, sometimes scabrous on the larger abaxial veins, the broadest blades with 6–9 conspicuous (dry) longitudinal nerves. Racemes 12–37 cm long, 3–16 cm wide, with a well-developed long central axis bearing numerous uncrowded thin, wiry, often-straight, spreading or ascending side branches 2–12 cm long. Spikelets ca. 2–3 mm long, bearing 1–3 florets on an elongate rachilla. Glumes 1.7–2.1 mm long, narrowly lanceolate or narrowly elliptic, uninerved, subequal, scabrous on the keel (and sometimes pubescent on the lamina), usually shorter than the spikelet. Lemmas 9–12 mm long, uninerved, sometimes pubescent marginally.

Field recognition characters.—Sheaths usually hirsute with glabrous blades. Racemes much longer than broad with long central axis and with numerous filiform spreading to ascending side branches.

Distribution and habitat.—A common weed in Barbados. The species ranges from the U.S. to South America.

Comment.—Pohl (1980) mentioned hybridization between this species and *L. virgata.*

Representative specimens: **BARBADOS.** No locality data. Common weed, "24.8.1901," *Bovell 268* (BAR); common weed, "15.9.1901," *Bovell 268* (BAR). No locality data nor date. *Herbarium, Dept. of Agriculture, Barbados 268* (BAR). Parish not specified. Hansen's Plantation, cultivated soil, weed, Feb 1937, *Allan 5a* (BAR). **St. George:** Cane Field, Rose Cottage [at Gun Hill], weed of cultivated ground, Jul 1937, *Allan 5a* (BAR). **St. John:** Codrington College, waste land near woods, Aug 1940, *Gooding 5* (BAR). **St. Michael:** Codrington Botanic Station, Nov 1935, *Mayers 5* (BAR). **St. Philip:** Canefield at base of Mt. Pleasant, 25 Sep 1996, *Rogers 96-63* (BAR, FLAS); Dodds, "24.8.01," *Bovell 268* (BAR).

40. **Leptochloa virgata** (L.) P. Beauv., Ess. Agrostogr. 71, 161, 166. 1812. (**Figs. 25, 26**).
BASIONYM: *Cynosurus virgatus* L., Syst. Nat. ed. 10, 2:876. 1759.

Culms becoming glossy and a little woody, sometimes glaucous on culms and foliage. Sheaths 5.5–10(–14) cm long, glabrous. Blades 12.5–30 cm long and 7–12 mm wide, the wider ones with up to 12 conspicuous longitudinal veins and the midrib strongly differentiated from the other longitudinal veins, widest below the middle, the apex long-attenuate. Inflorescences up to 19 cm long, to 17 cm wide, purplish (fresh and sometimes when dry), racemes with a long upright central axis bearing a fascicle of ascending filiform side branches 8–11 cm long. Spikelets 2–3 mm long, with (1–)5–8 florets. Glumes lanceolate, uninerved, scabrous on the thickened keels, the first glume about 3/4 the length of the spikelet or the glumes subequal, the second glume as long as the spikelet. Lemmas 1.7–1.8 mm long, lanceolate, awnless or sometimes with awns shorter than the rest of the lemma, sometimes fringed marginally.

Field recognition characters.—Large weedy grass with a purplish raceme bearing filiform side branches; florets usually 5 or more per spikelet; sheaths glabrous.

Distribution and habitat.—This species is found from the U.S. to South America and is scattered as a weed across Barbados.

Comments.—*Leptochloa virgata* differs from *L. mucronata* by having the lemmas over 1.7 mm (vs. under 1.2 mm long). In Barbados *L. virgata* tends toward glabrous sheaths as opposed to hirsute in *L. mucronata.* Additional tendencies are for *P. virgata* to have usually 5 or more vs. usually 1–3 florets per spikelet.

Representative specimens: **BARBADOS. Christ Church:** Canefield weed, Skeenes Hill, 9 Jan 1998, *Rogers 98-2* (BAR). **St. Andrew:** Entrance to Turner's Hall Wood, dry pasture, Jan 1939, *Allan 507* (BAR); Turner's Hall Woods, meadow in full sun, 17 Oct 1997, *Rogers 97-47* (BAR, FLAS). **St. Michael:** Weed on UWI campus grounds, 5 Nov 1997, *Rogers 97-50* (BAR, FLAS).

Lithachne P. Beauv., Ess. Agrostogr. 135. 1812.

Monoecious grasses with unisexual spikelets, the axillary inflorescences having a pistillate terminal spikelet surrounded by staminate spikelets. Lemmas and paleas of pistillate spikelets indurate, the glumes with long-attenuate apices.

A small genus of the New World warm climates.

Lithachne differs most conspicuously from the similar *Olyra* by having axillary as well as terminal inflorescences, with the pistillate flowers restricted to the tips of the axillary inflorescences. *Olyra*, by contrast, has the pistillate flowers mixed with staminate flowers in terminal panicles. *Lithachne* has a distinctive helmet-shaped lemma on the pistillate flowers (vs. ellipsoid lemmas on pistillate flowers in *Olyra*). Among our species, *Olyra* is larger-statured than *Lithachne*.

41. **Lithachne pauciflora** (Sw.) P. Beauv., Ess. Agrostr. 135, 168. 1812. BASIONYM: *Olyra pauciflora* Sw., Prodr. 21. 1788.

Shade-loving grass. Culms tough, woody, darkened at the internodes, these with a swollen collar above a ringlike groove. Sheaths 6–42 mm long, hispid (especially toward the leaf blade) to glabrous, auriculate at the collar, the pseudopetioles adaxially hispid. Blades lanceolate, 3.5–9.5 cm long, 8–23 mm wide, with acuminate apices and with obtuse or rounded asymmetric, pseudopetiolate bases, having up to 8 longitudinal nerves, both surfaces puberulent to glabrous, with pubescence near the base. Inflorescences up to ca. 5 cm long, paniculate or racemose. Terminal panicles staminate, on long wiry peduncles, with pedicellate spikelets. Axillary panicles (or racemes) with terminal pistillate spikelets and with lateral staminate spikelets. Staminate spikelets ca. 7 mm long, one-flowered; glumes not evident; lemma long and narrowly lanceolate

FIG. 25. *Leptochloa virgata* (*Rogers 98-2*)
FIG. 26. *Leptochloa virgata* (*Rogers 98-2*)

with an acuminate to acicular apex, trinerved, the palea similar and only a little smaller. Pistillate spikelets 12–14 mm long, one-flowered, with the subequal, narrowly lanceolate, multinerved glumes having long-attenuate acicular apices accounting for much of the length of the spikelet; lemma glossy white, indurate, obdeltoid (while folded), folded around a much smaller palea of similar texture.

Field recognition characters.—Broad-leaved shade-loving, monoecious grass with the terminal panicles strictly staminate, and with axillary panicles (or racemes) tipped by pistillate spikelets above staminate spikelets. Glumes of pistillate spikelets drawn out into long acicular apices (also true of *Olyra*).

Distribution and habitat.—This gully species occurs from Mexico to South America.

Representative specimens: **BARBADOS. St. James:** Kendal Gully, 25 Jul 1937, *McIntosh 410* (BAR). **St. Joseph:** Rusher Gully, 30 Jan 1998, *Rogers 98-17* (BAR, FLAS). **St. Thomas:** Welchman Hall Gully, 14 Aug 1996, *Rogers 96-3* (BAR).

Olyra L., Syst. Nat. ed. 10, 2:1261. 1759.

Monoecious bambusoid grasses with woody culms. Inflorescences paniculate, with pistillate spikelets toward the tips and staminate spikelets more basal. Pistillate spikelets with indurate lemmas. Staminate spikelets glumeless, not indurate.

A small genus of warm-climate America and of Africa.

42. **Olyra latifolia** L., Syst. Nat. ed. 10, 2:1253, 1261, 1379. 1759. (**Figs. 27, 28**).

Shade-loving grass. Culms rising to well over 1 m tall, hard, woody, having swollen, darkened nodes. Sheaths 4–8 cm long, auriculate at the collar, glabrous to hispid-strigose, with most pubescence apical-marginal. Blades 8–18 cm long, 1.9–4.0 cm wide, elliptic-oblong to lanceolate with acuminate tips and rounded asymmetric bases narrowed to pubescent pseudopetioles, with a conspicuous midnerve and ca. 7–9 pronounced lateral nerves; adaxial surface glabrous or

FIG. 27. *Olyra latifolia* (*Rogers 96-151*)
FIG. 28. *Olyra latifolia* (*Rogers 96-151*)

nearly so; abaxial surface glabrate with minute serrulations on major nerves; margins minutely serrulate. Inflorescences 6–11 cm tall, 2.5–4.5 cm wide, open, wiry, terminal panicles having ascending primary branches. Staminate spikelets ca. 7 mm long (including the awn), without glumes apparent; lemma narrowly lanceolate, trinerved, drawn out into an apical awn accounting for about half the length of the lemma; palea without an awn and about as long as the expanded portion of the lemma. Pistillate spikelets 16–27 mm long (including the awn), the first glume as long as the spikelet, ca. 9-nerved, pubescent toward the apex, lanceolate with a long-attenuate apex drawn out into an awn; second glume shorter, less pubescent, and fewer-nerved but otherwise similar; lemma much shorter then glumes, indurate, glossy, wrapped around the indurate palea.

Field recognition characters.—Broad-leaved, shade-loving, monoecious grass with staminate and pistillate florets in mixed terminal panicles (as opposed to being mixed in axillary inflorescences in *Lithachne*), the glumes on the pistillate spikelets drawn out into long acicular apices.

Distribution and habitat.—This attractive species is known in Barbados from Jack-in-the-Box Gully and Turner's Hall Woods, where it is widespread and abundant on the deeply shaded, comparatively moist forest floor. Its range extends from Mexico to South America.

Representative specimens: **BARBADOS. St. Andrew:** Turner's Hall Wood, forest floor, 1 Nov 1989, *Carrington 1087* (BAR); Turner's Hall Wood, damp soil in shade, 4 Oct 1937, *McIntosh & Allan 443* (BAR); Turner's Hall Woods, deep shade, on moist slope, 17 Dec 1996, *Rogers 96-151* (BAR, FLAS).

Oplismenus P. Beauv., Fl. Oware 2:14. 1810, nom. cons.

Sprawling or decumbent, broad-leaved grasses. Spikelets with a lower staminate or neuter floret and with an upper perfect floret. Glumes awned.

A small genus of warm climates.

KEY TO *OPLISMENUS* SPECIES IN BARBADOS

1. Awns scabrous .... **Oplismenus burmannii** P. Beauv., Ess. Agrostogr. 54. 1812.
   [An Asian species naturalized in the Neotropics. Reported by Gooding et al. from Rusher Gully and Foster Hall Woods; not encountered during the present study in either of these two locations, no materials in BAR.]
1. Awns smooth .... **Oplismenus hirtellus**

43. **Oplismenus hirtellus** (L.) P. Beauv., Ess. Agrostogr. 54, 168, 170. 1812. (**Figs. 29, 30**).

BASIONYM: *Panicum hirtellum* L., Syst. Nat. ed. 10, 2:870. 1759.

Culms sprawling, pubescent, rooting at the nodes. Sheaths 0.9–4.5 cm long, compressed, often with bilayered hirsute and puberulent pubescence, the hairs frequently pustule-based, some as long as 4 mm, or the sheath merely puberulent. Blades 2.5–9.5 cm long, 3–15 mm wide, lanceolate or lance-elliptic with an acuminate apex and with the asymmetric base acute, the midvein conspicuous, with 6–7 prominent lateral veins on the larger blades, often with bilayered pubescence on both surfaces, a puberulent layer beneath a hirsute, pustule-based layer, or merely puberulent. Panicles 4–10.5 cm long, 1–2 cm wide, the primary branches uncrowded, stiff, straight, and ascending, up to 16 mm long (excluding the awns), bristling with well-exposed, conspicuous awns. Spikelets 3.5–4.5 mm long (excluding awns) crowded. First glume lance-deltoid, pubescent, about half as long as the spikelet, armed with a near-apical, abaxial awn 10–12 m long, with paired short cusps rising above the insertion of the awn. Second glume longer than the first, its awn 2 mm long, the cusps less well-developed than those on the first glume. Sterile lemma lance-elliptic, pubescent, as long as the spikelet, its awn 1 mm long. Fertile lemma shorter than the sterile one, awnless, smooth, wrapped around the palea.

Field recognition characters.—Broad-leaved, shade-loving grass, heavily pubescent, often with bilayered, pustule-based hairs, especially on sheaths; spikelets with smooth awns on the glumes and on the sterile lemmas.

Distribution and habitat.—Distributed from Mexico to South America, *Oplismenus hirtellus* is the most abundant broad-leaved, shade-loving, forest understory grass in Barbados. It often forms substantial stands in such forest remnant areas as Turner's Hall Woods, the wooded area atop Cherry Tree Hill, and the forested area on and around Hackleton's Cliff.

Representative specimens: **BARBADOS. St. Andrew:** Turner's Hall Woods, Jan 1940, *Gooding 501* (BAR); Turner's Hall Woods, deep shade, 17 Dec 1996, *Rogers 96-155* (BAR, FLAS, K) and *96-156* (BAR); same locality, 27 Jan 1998, *Rogers 98-10* (BAR), *Rogers 98-12* (BAR). **St. Joseph:** Hackleton's Cliff, "5.1.18," *Herbarium, Royal Botanic Gardens, Trinidad & Tobago No. 8688* (BAR); Foster Hall Wood, damp shade, grazing for sheep, Dec 1937, *Allan 501* (BAR); Foster Hall Wood, damp shade, grazing for sheep, Dec 1939, *Allan 501* (BAR) [note the previous citation, which is identical except for the year]; Rusher Gully, 30 Jan 1998, *Rogers 98-16* (BAR).

Panicum L., Sp. Pl. 55. 1753.

Inflorescence a panicle. Spikelets with 2 florets, the lower floret neuter (or staminate) with a lemma resembling the glumes, and the upper one perfect and having an indurate lemma wrapped around the palea, this indurate and 2-nerved. Glumes both apparent and shorter than the spikelet.

A large, diverse, cosmopolitan genus. *Paspalum* differs from *Panicum* by having unbranched lateral branches in the inflorescence (vs. these often branched), spikelets especially strongly compressed (with the second glume and the lower lemma equal and resembling the valves of a clam), and the first glume usually absent.

KEY TO *PANICUM* SPECIES IN BARBADOS

1. Blades on pseudopetioles; spikelets hanging straight down from panicle branches; growing in shade on forest floor .... **Panicum pilosum**
1. Blades sessile; spikelets inserted on panicle branches at various angles; growing in sun in disturbed sites 2
   2. Grasses mostly over 1 m tall, with leaf blades often 1 cm wide; upper lemmas transversely rugose .... **Panicum maximum**
   2. Grasses generally shorter than 1 m, the leaf blades narrower than 0.6 cm; upper lemmas smooth 3
      3. Blades mostly ca. 3 mm wide; spikelets ca. 3 mm long .... **Panicum diffusum**
         [Rare if a component of the flora at all, see below.]
      3. Blades up to 14 mm wide; spikelets ca. 1.5 mm long .... **Panicum laxum**

FIG. 29. *Oplismenus hirtellus* (*Rogers 98-12*)
FIG. 30. *Oplismenus hirtellus* (*Rogers 98-12*)

44. **Panicum diffusum** Sw., Prodr. 23. 1788.

Culms slender, with dark, glabrous nodes. Sheaths up to 7 cm long. Blades to 25 cm long and 3(–4) mm wide, linear, nearly glabrous but with long pustule-based hairs marginally and sometimes adaxially toward the base, the conspicuous longitudinal nerves ca. 9 in the broadest leaf blades, the bases mostly tapered. Panicle ca. 14 cm long, 5.5 cm wide, open, airy, wiry, the axes sinuous. Spikelets ca. 3 mm long, uncrowded. First glume a little less than half the length of the spikelet, broadly ovate, mucronate. Second glume 7-nerved, lanceolate, the apex acuminate-cuspidate, as long as the spikelet. Lower lemma resembling the second glume, the lower palea well-developed and a little smaller than the lower lemma. Upper lemma and palea smaller than the lower ones, indurate, smooth.

Field recognition characters.—Leaf blades very narrow for their length (mostly under 3 mm wide while readily exceeding 20 cm long) and with a few long, white pustule-based cilia along the basal margins and at the collar; spikelets ca. 3 mm long.

Distribution and habitat.—This species occurs from the U.S. to South America. It has not been reported previously for Barbados, and it is scarcely documented as a component of the flora. The species is known in Barbados as just one specimen from a single now-extirpated clump on the University of the West Indies campus very near the biology buildings. Whether it is truly a member of the Barbados flora awaiting the discovery of new sites or an escape from imported biological materials is not clear.

Comments.—*Panicum diffusum* resembles *P. laxum* but has narrower leaves with tapered (vs. rounded) bases, long pustule-based cilia on the leaf margins, and far larger spikelets (ca. 3 mm vs. under 1.5 mm long). The identification of this collection as *P. diffusum* was challenging and is grounded in comparison with Antillean specimens from the Royal Botanic Gardens, Kew (K).

Representative specimens: **BARBADOS. St. Michael:** Weed on University of the West Indies Campus, growing at base of concrete wall along parking area, 4 Sep 1997, *Rogers 97-36* (BAR) [second collection from the same clump, 15 Sep 1997, *Rogers 97-42*].

45. **Panicum laxum** Sw., Prodr. 23. 1788.

Culms to ca. 80 cm tall, the lower portions sometimes decumbent and rooting at the nodes, these darkened. Sheaths to 10.5 cm long, glabrous except for a beard at the collar or with marginal hairs. Blades up to 20 cm long and 11 mm wide, glabrous except for scaberulous margins and sometimes abaxial major veins, the bases often auriculate (or merely truncate or rounded basally). Inflorescence ca. 24 X 10 cm, a delicate pyramid bearded or beardless at the major branching points and with puberulent axes. Spikelets 1–1.5 mm long, pedicellate. First glume ovate with an acute apex, usually about half the length of the spikelet, 3(–5)-nerved. Second glume lance-elliptic, as long as the spikelet, 3- or 5-nerved. Lower lemma as long as the spikelet, thin-textured, elliptic with an acute apex, 3- or 5-nerved, its palea almost as large as the lemma (though narrower). Upper lemma indurate, smooth, off-white.

Field recognition characters.—Slender grass with narrow blades having rounded bases; open panicles bearing spikelets under 1.5 mm long.

Distribution and habitat.—The only known population in Barbados is in Turner's Hall Woods, where the species is locally abundant on an exposed sandstone outcrop. The overall range extends from Mexico to South America.

Representative specimens: **BARBADOS.** No locality data, 26 Sep 1906, *Dash 268* (BAR). **St. Andrew:** Turner's Hall Woods, on the hilltop in a grassy, scrubby area in full sun, sandstone-based soil, 3 Dec 1998, *Rogers 98-80* (BAR).

46. **Panicum maximum** Jacq., Icon. Pl. Rar. 1:2. 1781. BASIONYM: *Urochloa maxima* (Jacq.) R. D. Webster, Austral. Paniceae 241. 1987 [1988]. [See comments under *Urochloa*.]

Large, imposing, pasture grass. Culms exceeding 2 m tall, bearded at the nodes. Sheaths 20 cm long, often hirsute with pustule-based hairs, the pubescence particularly concentrated at the collar, here with cilia sometimes as long as 6 mm. Blades to over 50 cm long and 13 mm wide, scabrous-margined, often with a prominent white (fresh and dry) midrib, sometimes hispid abaxially, especially toward the base. Panicles 14–32 cm long and 7–24 cm wide, pyramidal, open, wiry, with whorled, ascending, straight, often subbranched primary branches to 17 cm long, these devoid of spikelets near the base, otherwise bearing uncrowded spikelets on sinuous pedicels. Spikelets 3.2–3.5 mm long; first glume about 1/3 the length of the spikelet, broadly ovate with the base wrapped around the spikelet base, apically acute to truncate; second glume as long as the spikelet, lance-elliptic, 5-nerved. Lower lemma resembling the second glume, lanceolate, 5-nerved, bearing a staminate flower and with well-developed palea. Upper lemma and palea indurate, white, transversely rugose.

Field recognition characters.—Usually large (up to 2 m tall) with correspondingly large, open, pyramidal panicle; with long cilia at the collar; spikelets over 3 mm long, the upper lemma and palea rugose.

Common name.—Guinea Grass.

Distribution and habitat.—*Panicum maximum* is indigenous to Africa and is a widely cultivated forage species. This is one of the most abundant and most conspicuous grasses in Barbados. It is used for fodder, serves as a mulch, and is often the dominant grass along roadsides and in other disturbed sites.

Comments.—Specimens on the sandstone hills in the Scotland District of St. Andrew Parish tend to be smaller-statured than their lowland conspecifics.

Representative specimens: **BARBADOS.** Parish not specified. Waterford, from Pear Tree Field, fodder grass, "3.9.01," [one specimen also dated "Oct 4.11"], *Bovell 259* (BAR). **St. Andrew:** Baxters, sandstone hilltop, rocky, exposed, many specimens small-statured, spikelets with conspicuous fungal infection, 20 Oct 1997, *Rogers 97-49* (BAR). **St. George:** Gun Hill, Rose Cottage, moist pasture, cultivated for fodder, Dec 1939, *Allan 517* (BAR); in broken aging pavement around oil pump, just N of Highway 5 intersection with road to Turnpike & Carmichael, 16 Oct 1996, *Rogers 96-87* (BAR). **St. Michael:** Univ. of the West Indies campus, weed in unmowed area, 13 Oct 1997, *Rogers 97-46* (BAR).

47. **Panicum pilosum** Sw., Prodr. 22. 1788. (**Figs. 31, 32**).

Broad-leaved, shade-loving forest grasses. Culms to 70 cm tall but usually much shorter.

Fig. 31. *Panicum pilosum* (*Rogers 97-48*)
Fig. 32. *Panicum pilosum* (*Rogers 97-48*)

Sheaths and pseudopetioles 7.5–14 cm long (sometimes the entire sheath free of the stem and comprising the pseudopetiole), often with long marginal silky hairs, especially bearded at the junction between the blade and pseudopetiole. Blades 5–25.5 cm long, 7–16 mm wide, narrowly lanceolate, with a long-attenuate apex, serrulate-margined, adaxially glabrous or glabrate, abaxially glabrous or a little scabrous on the major veins, the base rounded or truncate. Panicles 9–21 cm long, 2.5–5.5 cm wide, narrowly pyramidal, the branches to ca. 5 cm long, inserted singly or in small clusters, unbranched (or minimally branched), straight, spreading or ascending, with the spikelets dangling vertically among long dangling cilia. Spikelets 1–1.5 mm long, short-petiolate, borne singly or less often in pairs. First glume broadly ovate when opened, about half the length of the spikelet, trinerved, acute apically. Second glume as long as the spikelet, lance-elliptic, 5-nerved. Lower lemma resembling the second glume, sterile, as long as the spikelet, strongly trinerved, its palea hyaline. Upper lemma and palea shorter then the lower lemma, bony white, indurate.

Field recognition characters.—Broad-leaved shade-loving grass with pyramidal inflorescences, these bearing spikelets hanging straight downward.

Distribution and habitat.—*Panicum pilosum* is restricted in Barbados to the moist, shaded forest floor of Turner's Hall Woods, where it is abundant. The nativity of this wide-ranging species is not clear.

Representative specimens: **BARBADOS. St. Andrew:** Turner's Hall Woods, damp undergrowth, Feb 1937, *Allan 466* (BAR); Turner's Hall Woods, scattered on deeply shaded spots, 17 Oct 1997, *Rogers 97-48* (BAR, FLAS).

Paspalidium Stapf, Fl. Trop. Afr. 9:582. 1920.

Grasses of wet habitats, having aerenchymatous culms. Inflorescence with erect, unbranched side branches along a long main axis. Spikelets with 2 florets, the upper one perfect, the lower one staminate or neuter. Lemma of fertile flower indurate.

A small pantropical genus.

48. Paspalidium geminatum (Forssk.) Stapf, Fl. Trop. Afr. 9:583. 1920. BASIONYM: *Panicum geminatum* Forssk., Fl. Aegypt.-Arab. 18. 1775.

Slightly thickened, stolon-forming grass. Culms 38 cm tall [specimen immature]. Sheaths loose and papery, as long as the blades. Blades 6–13 cm long, 3–6 mm wide, sometimes widest at or above the middle. Primary axes of the inflorescence ca. 0.5 cm long [immature], vertical, broad and flat, with the spikelets on one side. Spikelets 2–2.5 mm long. First glume half the length of the spikelet, as broad as long, truncate or obtuse. Second glume ovate, as long as the spikelet. Lower floret staminate, its lemma resembling the second glume and wrapped slightly around the lance-elliptic palea. Upper lemma ovate, thick-textured, wrapped around the palea.

Field recognition characters.—Grass of wet habitats; inflorescence with numerous vertical, straight, unbranched side branches.

Distribution and habitat.—The pantropical *Paspalidium geminatum* is known in Barbados from just one 1938 collection from Chancery Lane Swamp. Chancery Lane Swamp still exists as a seasonal marsh separated from the sea by the beach and low dunes. The surrounding area has been built up as suburban residential developments. Repeated field trips in different seasons to the "swamp" have failed to turn up the grass. Our description comes from the only collection.

Representative specimen: **BARBADOS. Christ Church:** Chancery Lane Swamp, in water, Jan 1938, *Allan 500* (BAR).

Paspalum L., Syst. Nat. ed. 10, 2:855. 1759.

Species tufted, or rhizomatous or stoloniferous. Culms 10–180 cm long, the nodes pubescent or glabrous. Primary inflorescence branches 1–many. Spikelets solitary or paired in 2–4 rows, sessile, subsessile or pedicellate, elliptic to suborbicular, compressed dorsiventrally, planoconvex. First glume usually absent. Second glume awnless, membranous. First lemma sterile, equal to second glume, awnless, membranous. Second lemma fertile, indurate or not, enclosing palea, this textured like its lemma.

KEY TO *PASPALUM* SPECIES IN BARBADOS

1. Spikelets with a broad, fimbriate wing .... **Paspalum fimbriatum**
1. Spikelets without a wing 2
  2. Inflorescence with spikes arising from an elongate central axis > 1.5 cm long 3
  2. Inflorescence with spikes radiating from a single point (or central axis < 1.5 cm long) 5
    3. Spikes 1–6; spikelets in 2 rows along rachis; leaf blades up to 55 cm long .... **Paspalum laxum** [and see comments on *P. setaceum* under *P. laxum*.]
    3. Spikes 4–22; spikelets often in 2–4 rows along rachis; leaf blades up to 85 cm long 4
      4. Spikelets pubescent, 2–3 mm long; spikes 4–11 in number; collar of lower leaves with an abaxial band of hairs (2–3 mm long) .... **Paspalum virgatum**
      4. Spikelets glabrous, 1.5–2 mm long; spikes 13–22 in number; collar of lower leaves glabrous abaxially .... **Paspalum millegrana**
      5. Spikes broadly divaricate at apex of culm, 2(–3) in number, 6–12 cm long; spikelets > 90 per spike, 1 mm long; leaf blades narrowly lanceolate, 0.5–1 cm wide .... **Paspalum conjugatum**
      5. Spikes erect or slightly spreading at apex of culm, 2–6 in number, 2.5–8 cm long; spikelets 24–50 per spike, 2–4 mm long; leaf blades linear, 0.2–0.5 cm wide 6
        6. Second glume pubescent; spikes often curved; fertile lemma usually bony at anthesis; second glume usually wider above the middle; one spike sessile and one pedunculate; freshwater habitats .... **Paspalum distichum**
        6. Second glume glabrous; spikes straight; fertile lemma leathery at anthesis; second glume usually wider at or below the middle; spikes pedunculate; saline habitats .... **Paspalum vaginatum**

49. Paspalum conjugatum Bergius, Acta. Helv. Phys.-Math. 7:129. 1762. (**Figs. 33, 34**).

Stoloniferous grass. Culms rising to ca. 50 cm, ca. 0.6 mm in diameter below the inflorescence, rooting at nodes, thin, glabrous. Sheaths 4–17 cm long, flattened, opened. Blades 2–16 cm long and 0.5–1 cm wide, glabrous except for few silky hairs at the collar, flat, elliptic. Inflorescences ca. 4.5–9 cm above leaves, consisting of 2(–3) thin, sessile, whiplike spikes of about equal length, these 6–13 cm long, bearing numerous spikelets arranged in 2 rows on one side of the rachis. Spikelets 1 mm long, light brown or green when fresh, borne on short pedicels. Second glume membranous, pubescent, ovate, acute at apex. Sterile lemma resembling upper

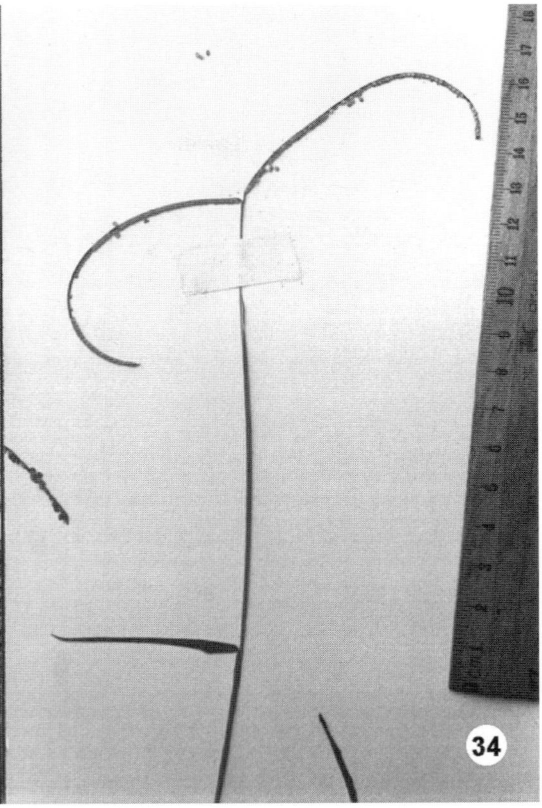

FIG. 33. *Paspalum conjugatum* (*Holder 97-14*)
FIG. 34. *Paspalum conjugatum* (*Holder 97-14*)

glume in shape, acute at apex, membranous, glabrous. Upper floret indurate, ovate, acute at apex, glabrous.

Field recognition characters.—Grasses of shaded, often moist places; inflorescences usually consisting of two long, thin, curved, whiplike spikes.

Distribution and habitat.—*Paspalum conjugatum* is a common weed of moist, grassy sites, and tolerates shaded conditions. Its range extends from the U.S. to South America, and it is naturalized in the Old World. In Barbados this species favors stream banks and shaded grassy areas. *Paspalum conjugatum* probably came to Barbados with livestock, although mature plants may cause digestive disturbances in cattle.

Representative specimens: **BARBADOS.** No locality specified. 4 Feb 1943, *Allan 269* (BAR). **St. Andrew:** Turner's Hall Woods, pasture just outside wood, Aug 1937, *Allan 509* (BAR); Turner's Hall Woods, meadow, 19 Jan 1997, *Rogers 97-6* (BAR); Chalky Mount, damp, shaded area at roadside, 23 Jun 1997, *Holder 97-6* (BAR, FLAS). **St. James:** Sunset Crest, shaded drainage ditch leading from golf course, 30 Jun 1997, *Holder 97-11* (BAR, FLAS). **St. John:** Pot House Stream, damp, shaded area on bank of stream, 1 Jul 1997, *Holder 97-14* (BAR). **St. Joseph:** Flower Forest, damp, shaded area at side of pathway, 20 Jun 1997, *Holder 97-5* (BAR, FLAS); weed in long, neglected lawn, Andromeda Gardens, 17 Jun 1997, *Rogers 97-26* (BAR). **St. Philip:** Dodds, weed, 13 Sep 1901, *Bovell 269* (BAR).

50. Paspalum distichum L., Syst. Nat. ed. 10, 2:855. 1759. (**Figs. 35, 36**).

Paspalum paspalodes (Michx.) Scribn., Mem. Torrey Bot. Club 5:29. 1894.

Extensively stoloniferous, nearly glabrous grass. Culms rising to ca. 0.5 m, ca. 1 mm in diameter below the inflorescence, rooting at the lower nodes. Sheaths flattened and 2.5–5 cm long. Blades 3–9 cm long, 3–5 mm wide, sometimes with silky hairs near the collar. Inflorescences consisting of 2(–3), curved to sometimes straight spikes 2–5 cm long, one sessile and one pedunculate, these bearing ca. 24–35 spikelets in 2 rows along one side of the rachis. Spikelets 2–3 mm long, elliptic-lanceolate. Second glume membranous, pubescent, elliptic to wider above

FIG. 35. Site for *Paspalum distichum* (*Holder 97-13*)
FIG. 36. *Paspalum distichum* (*Holder 97-13*)

the middle, with acute apex. Sterile lemma resembling the second glume in shape, membranous, glabrous. Upper floret with usually bony lemma enclosing the palea, the lemma elliptic with an apical tuft of hair.

Field recognition characters.—Freshwater-loving grass with usually two spikes, these usually curved, one sessile, the other pedunculate.

Distribution and habitat.—With its original range unclear, *Paspalum distichum* is found from the U.S. to South America and in the Old World. In Barbados this species is limited to freshwater habitats, sometimes with the stolons floating on the water surface.

Comments.—Application of the name *P. distichum* has a history of controversy, as is summarized in Rogers & Holder (1998). In short, the Linnaean type of *P. distichum* is reportedly a mixed collection of what most authors treat as *P. distichum* and as *P. vaginatum*. Bor (1970) noticed this and, as recounted in Guedes (1976), pinpointed a plant on the mixed sheet in a fashion that can be construed debatably as lectotypification. Authors adopting this potential lectotypification were compelled into a set of highly disruptive and confusing nomenclatural consequences. It became necessary to switch the name *P. distichum* to the species conventionally known as *P. vaginatum,* and to change the name of the species conventionally known as *P. distichum* to *P. paspalodes.* It was also necessary to cite P. *vaginatum* as a (nomenclatural) synonym under *P. distichum* even when the two extremely similar species are not regarded as conspecific. Guedes (1976) strongly took exception to this confusing circumstance and selected a different individual on the mixed Linnaean sheet to serve as (lecto- or neo-)type of *P. distichum*. The type chosen by Guedes (1976) preserved conventional usage of the names *P. distichum* and *P. vaginatum.* Guedes (1976) argued that Bor (1970) never formally designated a lectotype, and that Bor's putative lectotypification cannot be followed because it was not adequately specific about the individual plant to serve as type. Lacking access to the specimens and to the literature in question, we have conformed with recent consensus by following Guedes in applying the names in the conventional fashion.

Paspalum distichum differs from *P. vaginatum* in Barbados by having: a restriction to freshwater (vs. saline) habitats; pubescent (vs. glabrous) second glumes; the fertile lemma usually bony (vs. leathery) at anthesis; spikes often curved (vs. consistently straight); glumes usually widest above the middle (vs. usually widest below the middle); and one spike of the pair sessile (vs. all spikes pedunculate) (Rogers & Holder, 1998).

Representative specimens: **BARBADOS. St. John:** Bathland, near to roadside bridge, freshwater stream flowing towards sea, growing on stream banks and into water at stream margins, 1 Jul 1997, *Holder 97-13* (BAR, FLAS); Codrington College, freshwater stream flowing through shaded lawn, on the stream bank and in the water, 11 Jul 1997, *Holder 97-17* (BAR, FLAS). **St. Joseph:** Andrews Plantation, roadside gutter, Jun 1936, *Mayers 472* (BAR). **St. Philip:** Three Houses Stream (which flows all year), growing in water at stream margins with stolons extending across surface of water, 17 Jun 1997, *Holder 97-2* (BAR); Three Houses Stream (at the source), just opposite park, growing on stream banks and in water at stream margins, 1 Jul 1997, *Holder 97-12* (BAR, FLAS).

51. **Paspalum fimbriatum** Kunth in H. B. K., Nov. Gen. Sp. 1:93. 1816. (**Figs. 37, 38**).

Culms 13–68 cm long, 0.5–1 mm in diameter below the inflorescence, tufted, thin, glabrous, flattened. Sheaths 4–10 cm long, flattened, imbricate, closed. Blades 8–29 cm long and ca. 7 mm wide, broad, flat, marginally pubescent. Inflorescences 2–3 per plant, with central axis 4–6 cm long, borne 5–10 cm above the leaves, consisting of 2–6 spikes. Spikes 3–7 cm long, minutely pubescent basally, bearing ca. 30–50 spikelets arranged in 2 rows on one side of the rachis. Spikelets 3 mm long, purple and/or green. Second glume membranous, ovate, apiculate at apex, the margin with a broad, indurate, notched wing ca. 1 mm long. Sterile lemma resembling the second glume in shape and in possession of wing. Upper floret ovate, obtuse at the apex, not winged, indurate.

Field recognition character.—Fimbriate wing surrounding spikelet.

Distribution and habitat—*Paspalum fimbriatum* is a weed of dry, rocky roadsides, roadside ditches and waste grounds throughout the island. Its distribution is limited to the West Indies, to Central America, and northern South America.

Representative specimens: **BARBADOS.** Without locality data. 23 Sep 1901, *Bovell 200* (BAR); Jun 1901, *Bovell 200* (BAR); 19 Sep 1906, *Dash 200* (BAR). [Note same "species number" for two different Bovell collections.] **St. Andrew:** Chalky Mount, roadside, full sun, 23 Jun 1997, *Holder 97-7* (BAR). **St. John:** Pool Land, rocky area at roadside, 18 Jun 1997, *Holder 97-4* (BAR); Bathland, at roadside, 1 Jul 1997, *Holder 97-15* (BAR). **St. Joseph:** Hackleton's Cliff and Bathsheba, 5 Jan 1918, *Herbarium-Royal Botanic Gardens, Trinidad and Tobago 8682* (BAR). **St. Michael:** Codrington House, waste ground, wayside, Dec 1936, *Allan 314* (BAR); UWI campus weed, full sun, highly disturbed ground, 20 Aug 1996, *Rogers 96-10* (BAR, FLAS).

52. **Paspalum laxum** Lam., Tabl. Encycl. 1:176. 1791. (**Fig. 39**).

Tufted grass. Culms 33–72 cm long and ca. 0.5 mm in diameter below the inflorescence, thin, wiry, glabrous, the nodes dark and glabrous. Sheaths 3–20 cm long, imbricate, pubescent at margins. Blades 6–55 cm long, 2–7 mm wide, puberulent or glabrous, linear. Collars pubescent. Inflorescences terminal and infrequently axillary, with central axis 2.5–11 cm long, consisting of 1–6 thin, usually curved spikes up to 9 cm long, each spike bearing numerous paired spikelets in 2 rows on one side of the basally pubescent rachis, this kinked at the apex. Spikelets up to 2 mm long, elliptic-ovate, acute at apex. Second glume glabrous or pubescent, elliptic or ovate, acute at apex, membranous. Sterile lemma glabrous, otherwise resembling the second glume. Upper floret indurate, glabrous, obovate to ovate, obtuse at apex.

Field recognition characters.—Lax, tufted grass with long (to 55 cm), narrow (2–7 mm wide) leaf blades widest near the middle; inflorescence with 1–6 thin, long (to 9 cm) spikes.

Distribution and habitat.—*Paspalum laxum* occurs in scattered habitats, favoring the eastern side of Barbados. Its distribution is limited to the West Indies and southern Florida Keys. Habitats include chinks in limestone rocks and various likewise alkaline sandy or clay-soiled seashore sites. It apparently has a preference for limestone (Hitchcock, 1950; Correll & Correll, 1982).

Comments.—This species closely resembles the highly polymorphic and poorly defined *P. setaceum*

FIG. 37. *Paspalum fimbriatum* (Weed at Chemistry Bldg. UWI campus)
FIG. 38. *Paspalum fimbriatum* (*Holder 97-15*)

Michx., which has a lengthy synonymy as well as multiple varieties (see Banks, 1966). Gould (1979) listed *Paspalum setaceum* as present in Barbados with no substantiation nor with specimens cited. Gould may have based this on a reference without specimen citations to *Paspalum propinquum* Nash in Gooding et al. (1965), this in turn based on Hitchcock (1936), and this based on the collection *Hitchcock 16500* (not seen). (Or Gould may have seen the Hitchcock specimen directly.) Gould (1979) as well as Banks (1966), Correll and Correll (1982), and others interpreted *P. propinquum* as a synonym of *P. setaceum*. Different authors separate *P. laxum* from *P. setaceum* differently and often not convincingly. Though inconsistent, characteristics of *P. setaceum* tend to include glandular hairs in the inflorescence (Hall, 1978), axillary inflorescences (these are occasional in *P. laxum* too), and differences in leaf morphology. Having trouble distinguishing *P. laxum* from *P. setaceum*, GR studied these species at the University of Florida Herbarium and found leaf morphology to supply the most consistent distinguishing characters for the materials previously identified as one species or the other at FLAS. Specimens believed to pertain to *Paspalum setaceum* showed enormous variation in the leaf blades, yet the blades are almost consistently broadest at or near the base, whereas specimens of *P. laxum* have leaf blades broadest near the middle and tapering toward the base. Further, *P. laxum* specimens tend toward longer, proportionately narrower leaf blades up to 55 cm long and only 5(–7) mm wide. Those of *P. setaceum*, by contrast, seldom have blades exceeding 30 cm long, with the longest ones 8–13 mm wide. All Barbados materials seen by us in the field and in BAR have blades characteristic of *P. laxum* and lack glandular hairs. Occasional axillary inflorescences are seen. It must be emphasized that our species concept is based on examination of herbarium specimens identified by others without examination of types.

Representative specimens: **BARBADOS. Christ Church:** Chancery Lane, seashore, generally in shade, Jan 1938, *Allan 502* (BAR); Chancery Lane, 16 Jul 1906, *Dash 584* (BAR); on dry disturbed ground, Graeme Hall Swamp, 3

FIG. 39. *Paspalum laxum* (*Holder 97-18*)

May 1999, *Rogers 99-54* (BAR). **St. Andrew:** Beach south of Belleplaine and north of Barclays Park, 18 Oct 1996, *Rogers 96-89* (BAR). **St. John:** Bath, in seaside woods among whitewoods, not abundant, 3 May 1997, *Rogers 97-18* (BAR); same locality, 11 Jul 1997, *Holder 97-18* (BAR); same locality, 16 Jan 1998, *Rogers 98-3* (BAR). **St. Joseph:** Tenby Bay House, 4 Nov 1901, *Freeurou & Bovell 584* (BAR); Bathsheba, grassland, limestone rocks, 7 Jan 1918, *West 8686* (BAR). **St. Philip:** Three Houses Park, on limestone rocks, full sun, 6 Dec 1996, *Rogers 96-148* (BAR, FLAS); Three Houses Park, on limestone rocks, 17 Jun 1997, *Holder 97-3* (BAR, FLAS).

53. Paspalum millegrana Schrad., Mant. 2:175. 1824. (**Figs. 40, 41**).

Large tufted grass. Culms up to ca. 140 cm tall, 1–2.5 mm in diameter below the inflorescence, glabrous. Sheaths 0.5–1 cm wide, 21–32 cm long, flattened, glabrous; collar pubescent-puberulent. Blades 31–85 cm long, 1–1.5 cm wide, lanceolate, pubescent on adaxial surface. Inflorescence with a central axis 10–16 cm long and bearing 13–22 spikes 7–21 cm long. Spikelets 1.5–2 mm long, paired in 2–4 rows on one side of the pubescent rachis, broadly obovate (or orbicular or elliptic), acute at the apex, green and/or purple (fresh and sometimes when dry). Second glume glabrous, broadly obovate (or orbicular or elliptic), acute at apex, membranous. Sterile lemma resembling the upper glume. Upper floret glabrous, obovate, acute at apex, indurate.

Field recognition characters.—Large grass on sandstone substrates (in Barbados), the leaf blades adaxially pubescent; central axis of the raceme over 10 cm long with equally long spikes; spikelets paired and glabrous.

Distribution and habitat.—A species of the West Indies and of Central and South America, *Paspalum millegrana* is unusual in Barbados, being restricted to exposed sandstone areas in the Scotland District of St. Andrew Parish, where the soil pH is uncharacteristically low for otherwise coral-derived Barbados (Vernon & Carroll 1965). At the Baxters sandstone outcrop *Paspalum millegrana* occurs in a mixed population with *Paspalum virgatum* which it closely resembles, as well as with *Cyperus ligularis* L., *Andropogon glomeratus,* and *Sporobolus indicus.*

Representative specimens: **BARBADOS.** No data. *Barbados Museum and Historical Society (BMHS) 700* (BAR). **St. Andrew:** Chalky Mount, 8 Aug 1904, *Murphy 455* (BAR); Baxters, exposed sandstone, 30 Jul 1997, *Holder 97-21* (BAR, FLAS); N of Chalky Mount Pottery, on exposed sandstone, 30 Jul 1997, *Holder 97-22* (BAR, FLAS).

54. Paspalum vaginatum Sw., Prodr. 21. 1788. (**Figs. 42, 43**).

Extensively stoloniferous grass. Culms infrequently ascending to ca. 0.5 m, ca.1 mm in diameter below the inflorescence, rooting at the nodes, glabrous to lightly pubescent. Sheaths 4–7 cm long, imbricate, flattened and opened (not fused) for most of their length. Ligule a little over 0.5 mm long, membranous, directly in front of a row of cilia many times the length of the ligule (cilia often 3–4 mm long). Blades 4–12 cm long, 2–3 mm wide, nearly glabrous, adaxially with uniform papillae (these rarely forming a minute puberulence). Inflorescence consisting of 2–6 spikes borne on short peduncles, the spikes straight, 2.5–8 cm long, bearing ca. 34–50 spikelets arranged in paired rows along either side of the rachis. Spikelets 2–4 mm long, ovate-elliptic. Second glume glabrous, elliptic to a little wider at or below the middle, acute at apex, membranous. Sterile lemma resembling the second glume in shape, glabrous or with some apical hairs, membranous. Fertile lemma elliptic, leathery, glabrous.

Field recognition characters.—Stoloniferous grass found on beach sand or in wet saline places; short-pedunculate spikes usually 2 or 3 (or occasionally more) and straight. Similar vegetatively to *Sporobolus virginicus* (see below).

Distribution and habitat.—Encountered around the world in warm coastal regions. In Barbados *P. vaginatum* occurs in two different habitats, both saline: as a sand-binding species on beaches, and in saline standing water (*Holder 97-1, 97-9*). Rogers and Holder (1998) found the salinity of the standing-water sites in Graeme Hall Swamp on the South Coast of Barbados to be 4–6 parts/1000 NaCl.

Comments.—The close resemblance between *P. vaginatum* and *P. distichum* is discussed under the latter species and in more depth in Rogers and Holder (1998). Vegetatively, *P. vaginatum* is almost identical to *Sporobolus virginicus*, and the two occur with their rhizomes snaking aggressively across the sand on beach habitats. In flower, the two differ substantially, with the *Sporobolus* producing a single, dense, whitish panicle in contrast with the paired (or multiple) digitate spikes of *Paspalum*. As examined at 40× magnification, the two species differ in that *P. vaginatum* has uniformly sized papillae (as opposed to dimorphic papillae) adaxially on the leaf blades, smooth (vs. usually minutely serrulate) blade margins, and long cilia directly behind the membranous ligule (vs. the ligule being made up of minute cilia with no row of long cilia behind). The adaxial foliar papillae on *Sporobolus virginicus* are of two conspicuously different lengths and of different morphologies: one type is low and rounded, and the second type is far longer and is

FIG. 40. Mixed population of *Paspalum millegrana* and *P. virgatum* at Baxters, St. Andrew.
FIG. 41. *Paspalum millegrana* (*Murphy 455*)

sharp-pointed, reminiscent of a rhinoceros horn. These large papillae usually occur on the blade margin of *S. virginicus,* giving it a serrulate-scabrous aspect. The ligule of *S. virginicus* is a band of tightly appressed hairs. Although long cilia do often occur around the collar in *S. virginicus*, the cilia are either marginal on the sheath or adaxial on the blade, forming a long silky overstory above the dimorphic papillae; the long cilia are not directly behind the ligule.

Representative specimens: **BARBADOS. Christ Church:** Graeme Hall Swamp, 11 May 1997, *Rogers 97-27* (BAR, FLAS); Graeme Hall Swamp, interior of swamp, adjacent to red mangroves, growing across water surface of muddy ditch, 11 Jun 1997, *Holder 97-1* (BAR); 25 Jun 1997, *Holder 97-9* (BAR, FLAS); Graeme Hall Swamp, to the west of South Coast Road entrance, adjacent to white mangroves, growing in wet substrate, *Holder 97-8* (BAR, FLAS); Casuarina Beach, on the sand, 11 Jul 1997, *Holder 97-16* (BAR, FLAS). **St. Andrew:** On beach, sprawling across sand just N of Barclays Park, 21 Aug 1996, *Rogers 96-23* (BAR); Beach S of Belleplaine and N of Barclays Park, 18 Oct 1996, *Rogers 96-90* (BAR). **St. Philip:** Crane Beach, 27 Sep 1901, *Bovell 277* (BAR).

55. **Paspalum virgatum** L., Syst. Nat. ed. 10, 2:855. 1759. (**Figs. 44, 45**).

Culms to 180 cm tall, ca. 2 mm in diameter below the inflorescence, tufted, glabrous. Sheaths ca. 24 cm long and 7 mm wide, imbricate, pubescent at margins; collar pubescent, on lower leaves with an abaxial band of white hairs 2–3 mm long. Blades 23–63 cm long, 1.5–2.3 cm wide. Inflorescences with central axis 12–17 cm long, consisting of 4–11 spikes 5–18 cm long, these with tufts of long silky hair at their bases, and bearing numerous paired spikelets sometimes arranged in 2–4 rows on one side of the pubescent rachis. Spikelets 2–3 mm long, broadly obovate (or orbicular or elliptic). Second glume densely pubescent, membranous, broadly obovate (or orbicular or elliptic), apiculate at the apex. Sterile lemma resembling the second glume in shape, membranous, apiculate and pubescent at the apex. Upper floret glabrous, obovate-elliptic, acute at the apex, indurate.

Field recognition characters.—Large grass of sandstone substrates (in Barbados); branches of the raceme with paired pubescent spikelets on one side of the rachis.

FIG. 42. *Paspalum vaginatum* (*Holder 96-17*)
FIG. 43. *Paspalum vaginatum* (*Holder 96-17*)

Distribution and habitat.—This species ranges from the southern U.S. to Brazil, including the West Indies. During the present study *P. virgatum* turned up at two localities in St. Andrew Parish. The site at Baxters was exposed sandstone where *Paspalum virgatum* was mixed with *P. millegrana* (see this species for comments on soil). The other site is in a depression where particles from the weathering of exposed sandstone at higher elevations would reasonably be expected to collect.

Comments.—*Paspalum millegrana* differs from the similarly large-statured *P. virgatum* by having glabrous spikelets 1.5–2 mm long (vs. spikelets pubescent and 2–3 mm long), over 13 (vs. under 11) spikes per inflorescence, and the collar glabrous (vs. pubescent) abaxially.

Representative specimens: **BARBADOS. St. Andrew:** Near Chalky Mount [Bruce Vale], Highway 2 between Baxters and Haggatts, roadside weed, 11 Sep 1996, *Rogers 96-46* (BAR, FLAS); same locality, at roadside, and spreading through adjacent waste ground, abundant, 30 Jul 1997, *Holder 97-19* (BAR, FLAS); Baxters on exposed sandstone, 30 Jul 1997, *Holder 97-20* (BAR).

Pennisetum Rich., Syn. Pl. 1:72. 1805.

Panicle compact and elongate with bristles among the spikelets (spikelets themselves with no awns). Spikelets 2-flowered, the upper floret perfect, the lower one staminate or neuter. First glume reduced.

In warm climates around the globe. *Pennisetum* differs from *Setaria* by having the bristles disarticulate with the spikelets.

KEY TO *PENNISETUM* SPECIES IN BARBADOS

1. Blades over 1 cm wide; grass to 3 m tall; inflorescence bristles antrorse-scabrous but not ciliate .... **Pennisetum purpureum**
1. Blades narrower than 1 cm; grass not exceeding 1.5 m tall; inflorescence bristles antrorse-scabrous and ciliate 2

FIG. 44. *Paspalum virgatum* with A. Holder (*Holder 97-19*)
FIG. 45. *Paspalum virgatum* (*Holder 97-19*)

2. Inflorescence axis glabrous or puberulent; spikelets ca. 4 mm long; bases of bristles formed into a false involucre by overlapping, interlocked pectinate cilia; plants uncultivated .... **Pennisetum setosum**
2. Inflorescence axis hispidulous; spikelets 5 mm long or longer; bases of bristles ciliate but the cilia not pectinate nor forming a false involucre; plants cultivated and infrequently escaped .... **Pennisetum setaceum**

[Highly variable and represented by multiple apparent cultivars. At least one cultivar has purple inflorescences and flat purple leaves. Others have flat to folded green leaves and near-white to purplish inflorescences. Fig. 46.]

56. **Pennisetum purpureum** Schumach., Beskr. Guin. Pl. 44. 1827. (**Fig.** 47).

Very large upright grass. Culms to ca. 3 m tall. Sheaths 10–14(or more) cm long, glabrous. Ligules mostly ca. 4 mm long, a band of stiff, dark-colored trichomes. Blades to ca. 70 cm long, ca. 2 cm wide, the broadest blades with about 10 conspicuous longitudinal nerves, sharply scabrous-margined, very long-attenuate apically, adaxially with long pustule-based trichomes toward the base, glabrous distally; abaxially glabrous except for serrulations on the midrib. Panicles 16–30 cm long, 2–3 cm wide, finger-shaped, tawny, bristly, compact and congested, vertical and continuing the axis of the culm, the axis densely covered with bilayered pubescence, the longer layer hirsute. Spikelets (excluding bristles) ca. 4 mm long, glabrous and free of awns, subtended by and dehiscing along with tufts of tawny, antrorsely barbed bristles up to ca. 13 mm long. First glume less than 1/3 the length of the spikelet (sometimes far less), single-nerved. Second glume less than 1/2 the length of the spikelet, lanceolate, delicate, trinerved; lower lemma lanceolate, essentially as long as the spikelet, prominently 3- or 5-nerved, micro-scaberulate, especially on the nerves, its palea hyaline. Upper lemma similar to and as large as the lower lemma but finer-textured, trinerved. Upper palea almost as large as the upper lemma.

Field recognition characters.—The fingerlike panicles are over 16 cm long, tawny, and bristly despite having awnless spikelets; the main axis of the panicle is densely hirsute.

Common name.—Elephant Grass.

FIG. 46. *Pennisetum setaceum* as an ornamental, Three Houses Park, St. Philip.
FIG. 47. *Pennisetum purpureum* as a weed, Andromeda Botanic Gardens, St. Joseph.

Distribution and habitat.—*Pennisetum purpureum* is encountered towering above Barbados roadsides. It prefers low, comparatively moist habitats. The species is indigenous to Africa, and is used as forage grass in many tropical regions.

Representative specimens: **BARBADOS. St. John:** "in Tenantry," Coach Hill, Nov 1935, *Mayers 14* (BAR). **St. Joseph:** Horse Hill, near the top (near turnoff for Cotton Tower), scattered along road in sunny weedy places, to ca. 3 m tall, 21 Aug 1996, *Rogers 96-18* (BAR, FLAS).

57. Pennisetum setosum (Sw.) Rich., Syn. Pl. 1:72. 1805. BASIONYM: *Cenchrus setosus* Sw., Prodr. 26. 1788.

Culms to 1.5 m tall. Sheaths 8.5–13 cm long, hirsute (especially marginally-apically) to glabrous. Ligules ca. 2 mm long, a row of light-colored hairs. Blades 12.5–34 cm long, 2–13 mm wide, with long-attenuate apices, scabrous-margined, adaxially usually hirsute, abaxially glabrous to infrequently hirsute. Inflorescences 8–20 cm long, 2–3 cm wide, finger-shaped, tightly congested, bristly, purplish (fresh and usually so when dry), the main axis scabrous or minutely puberulent. Spikelets subtended by and dehiscing with bristles of highly heterogeneous lengths up to 17 mm long, these featherlike basally with silky cilia, antrorsely barbed apically. Spikelets ca. 4 cm long. First glume a small inconspicuous scale ca. 1/5 the length of the spikelet. Second glume as long as the spikelet, delicate-textured, 5-nerved, lanceolate. Lower lemma similar to and about as long as the second glume but even more delicate, trinerved, provided with a thin palea. Upper lemma thickened, white, glossy, much shorter than the lower lemma.

Field recognition characters.—Weedy grass with purplish, bristly, cylindric, tightly congested panicles having the main axis scabrous or minutely puberulent.

Distribution and habitat.—*Pennistum setosum* extends from Florida to Brazil, including the West Indies. In Barbados it is a roadside weed.

Representative specimens: **BARBADOS.** No data. *Botanic Station Herbarium, Barbados 346* (BAR). No locality data. Roadside, 8 Oct 1906, *Dash 346* (BAR). **St. George:** Gun Hill, moist waste ground, Feb 1937, *Allan 468a* (BAR). **St. John:** Montcrieffe, gully side, Jan 1940, *Gooding 468* (BAR). **St. Joseph:** Bathsheba, roadside, 13 Oct 1989, *Carrington 1089* (BAR); Tenby Bay House, 30 Oct 1901, *Bovell & Freeman 346* (BAR). **St. Joseph/St. John:** Malvern Lodge Gully, 31 May 1902, *Freeman 346* (BAR). **St. Peter:** Highway 2 near Speightstown, roadside, 1.2 km S of Highway 1 intersection, at brink of gully, to 1.5 m tall, 4 Sep 1996, *Rogers 96-26* (BAR); pasture, Maynards, Jan 1936, *Mayers 468* (BAR).

Pharus P. Browne, Civ. Nat. Hist. Jamaica 344, t. 38, f. 3. 1756.

Shade-loving monoecious grasses with long (pseudo)petioles supporting the broad blades. Inflorescences long-pedunculate panicles with the spikelets uncrowded on stiff straight branches. Spikelets unisexual and one-flowered, in pairs with the pistillate spikelets sessile, and with the staminate spikelets smaller and pedicellate.

A small genus restricted to warm climates in the New World.

KEY TO *PHARUS* SPECIES IN BARBADOS

1. Blades to 5 cm wide, not glossy (fresh); lemmas of pistillate spikelets covered evenly with pubescence; glumes on pistillate spikelets covering less than 1/2 of the lemma .... **Pharus lappulaceus**
1. Blades to 7 cm wide, glossy (fresh); lemmas of pistillate spikelets pubescent at the apex only; glumes on pistillate spikelets covering more than 1/2 of the lemma .... **Pharus latifolius**

58. Pharus lappulaceus Aubl., Hist. Pl. Guiane 2:859. 1775. (**Fig. 48**).

Pharus glaber Kunth, Nov. Gen. Sp. 1:196. 1816. (See Webster & Peterson, 1996.)

Culms to 68 cm tall, the upper portion a slender, wandlike, puberulent peduncle rising far above the foliage. Sheaths up to ca. 12.5 cm long, glabrous, extensively overlapped. Blades (6–)16–21(–25) cm long, (1.5–)3.5–4.5(–5) cm wide, narrowly oblanceolate to elliptic, the apex acute to acuminate, sometimes with a drip-tip, the base lopsided and acute to acuminate, glabrous or glabrate, with a conspicuous midvein and with ca. 6–8 conspicuous lateral nerves and with innumerable fine lateral nerves between the more prominent ones, on pseudopetioles (1–)2–3 cm long. Stiff panicles to ca. 21 cm long and 15 cm wide, on a long wandlike peduncle, the axes hispidulous-tomentose, the primary branches to 13 cm long, straight and horizontal to ascending. Staminate (pedicellate) spikelets usually ca. 4 mm long, the glumes puberulent, the first glume ca.1/4 the length of the spikelet, lanceolate, the second glume similar but reaching1/2–3/4 the length of the spikelet, the lemma elliptic-lanceolate, multinerved, minutely puberulent to glabrate, as long as the spikelet. Pistillate (sessile) spikelets to 12 mm long, the lance-elliptic, multinerved glumes ca. 5–6 mm long, nearly equal, covering less than half of the lemma, the first glume broader than the second glume; lemmas to ca. 12 mm long, the most conspicuous components of the spikelet, covered densely on the abaxial surface with a thick tomentum of hooked hairs, becoming thick and hard, wrapped tightly around an equally long but narrowly linear palea.

FIG. 48. *Pharus lappulaceus* (*Rogers 96-152*)

Field recognition characters.—Broad, frequently narrowly oblanceolate leaf blades to ca. 5 cm wide showing cross-veins perpendicular to the longitudinal veins, on pseudopetioles over 1 cm long. Pistillate spikelets with lemmas often over 1 cm long, these less than half-covered by the glumes, tomentose with hooked hairs over their entire surface, and with the margins tightly inrolled.

Distribution and habitat.—*Pharus lappulaceus* is a striking grass of comparatively moist, shaded forest floors ranging from Florida, U.S., and Mexico to South America. In Barbados it is recorded only from Turner's Hall Woods, where this eye-catching grass has been collected several times over the years.

Comments.—A suite of unusual characteristics makes this species easy to recognize. The leaves are unusually broad for a grass, often exceeding 3.5 cm wide; they are often oblanceolate (with corresponding divergent lateral nerves), have cross-veins perpendicular to the longitudinal veins, and are long-pseudopetiolate. The pseudopetioles and overlapping, winged sheaths are reminiscent of members of the Zingiberales. The inflorescences are on long wandlike peduncles and bear uncrowded, stiff straight primary and secondary branches with uncrowded spikelets in pairs, one pistillate and sessile, the other staminate and stalked. The hardened cigar-shaped lemmas, ca. 1 cm long, protrude from the pistillate spikelets and are covered with hooked hairs, apparently an adaptation for dispersal.

Representative specimens: **BARBADOS. St. Andrew:** Turner's Hall Woods, deep shade, on moist slopes, 17 Dec 1996, *Rogers 96-152* (BAR, FLAS); same locality, 4 Jun 1998, *Rogers 98-40* (BAR); Turner's Hall Woods, herb of forest floor, 13 Mar 1989, *Carrington 874* (BAR); Turner's Hall Woods, moist shade, Oct 1937, *McIntosh & Allan 447* (BAR) [Note: One specimen labeled "*McIntosh & Allan "447=447a*" collected in Oct 1937 is *Pharus latifolius*.]; Turner's Hall Woods, floor of woods, Aug 1940, *Gooding 447* (BAR); Turner's Hall Woods, 26 Sep 1906, *Dash 624* (BAR).

59. **Pharus latifolius** L., Syst. Nat. ed 10, 2:1269. 1759.

Broad-leaved shade-loving grass. Culms to ca. 54 cm tall, wandlike, pubescent. Pseudopetioles to 10 cm long and not sharply differentiated from the sheaths. Blades to ca. 20 cm long, 4–7 cm wide, elliptic to oblanceolate, then often with a narrowly attenuate base, the apex with a drip-tip, having divergent lateral nerves. Panicles to 20 cm long and 7 cm wide, the primary branches to 15 cm long. Pistillate spikelets ca. 15–16 mm long, the glumes elliptic-lanceolate, multinerved, subequal, covering over 1/2 of the lemma, this usually with the pubescence concentrated at the apex. Staminate spikelets with the first glume 1/4 the length of the spikelet and the second glume 1/2 the length of the spikelet, the glumes lanceolate; lemma lanceolate.

Field recognition characters.—Elliptic to oblanceolate blades over 4 cm wide with pseudopetioles.

Distribution and habitat.—Occurring in Central America, South America, and the Caribbean, *Pharus latifolius* is known in Barbados only from Turner's Hall Woods.

Representative specimens: **BARBADOS. St. Andrew:** Turner's Hall Wood, moist shade, weed, Oct 1937, *McIntosh & Allan "447=447a,"* (BAR). [Note: different material with the collection number *"McIntosh & Allan 447"* collected in Oct 1937 is *Pharus lappulaceus*.]; same locality, 27 Jan 1998, *Rogers 98-11* (BAR), same locality, 4 Jun 1998, *Rogers 98-41, 98-43* (BAR).

Rhynchelytrum Nees, Staz. Sperim. Agrar. Ital. 378, 446. 1836.

Inflorescences paniculate, the spikelets 2-flowered. Spikelets silky pubescent. First glume reduced; second glume larger; lower lemma similar in size to second glume; fertile lemma much shorter than spikelet, smooth.

A small African genus.

60. **Rhynchelytrum repens** (Willd.) C. E. Hubb., Bull. Misc. Inform. Kew 1934:110. 1934. **(Fig. 49)**. BASIONYM: *Saccharum repens* Willd., Sp. Pl. Ed. 4, 1:322. 1797.

Tufted weedy grass. Culms to ca. 75 cm tall. Sheaths 8–18 cm long, sometimes hirsute (especially marginally). Ligule a fringe of hairs about 1 mm long. Blades 8–17 cm long, (1–)2–4 mm wide, linear, glabrous. Panicles 11–14 cm long, 3–4.5 cm wide, pyramidal, purplish and delicately feathery, the major branches very thin, ascending, kinked, and rebranched. Spikelets ca. 4 mm long (excluding the hairs), on sinuous pedicels. First glume a tiny pubescent scale. Second glume as long as the spikelet, densely silky-pubescent, lanceolate, bearing a subapical awn. Lower lemma resembling the second glume, similarly pubescent, lanceolate, awned. Upper lemma and palea much smaller than the lower lemma, delicate, glabrous, white.

Field recognition characters.—The feathery purplish panicle is unique.

Common name.—Natal Grass.

Distribution and habitat.—*Rhynchelytrum repens* is abundant and conspicuous in Barbados. This African grass flourishes on beaches, roadsides, and rocky places as well as in disturbed sites. In places it fills entire meadows.

Representative specimens: **BARBADOS. Christ Church:** Streetside weed at 135 Atlantic Shores, Oct 1996, *Rogers 96-99* (BAR). **St. Andrew:** Stream crossing near Haggats, 21 Aug 1996, *Rogers 96-15* (BAR). **St. Michael:** Codrington Botanic Station, Nov 1935, *Mayers 12* (BAR). **St. Peter:** Speightstown, on the beach, 29 Oct 1996, *Rogers 96-121* (BAR).

Rottboellia L. f., Suppl. Pl. 13, 114. 1782, nom. cons.

Large coarse grasses. Racemes highly modified into tight cylinders, the paired spikelets on a thickened, disarticulating rachis. In each pair, one spikelet sessile, 2-flowered, and sunken; the other on a broad, strap-shaped pedicel and appressed to the rachis.

Small genus indigenous to the Old World tropics.

61. **Rottboellia cochinchinensis** (Lour.) Clayton, Kew Bull. 35:817. 1981. **(Figs. 50, 51)**. BASIONYM: *Stegosia cochinchinensis* Lour., Fl. Cochinch. 1:51. 1790.

Large erect grasses. Culms 2–3 m or more tall, attaining 1 cm in diameter basally. Sheaths to 10 cm long, loose, often hirsute with stiff, brittle, patent, pustule-based hairs to 2 mm long. Ligule ca. 0.5 mm long, a band of hairs. Blades 10–47 cm long, 5–20(–28) mm wide, roughly scabrous-margined, with a conspicuous white (fresh and dry) midrib flanked by up to ca. 12(–14) longitudinal veins, mostly glabrous abaxially, often strigose adaxially. Inflorescences 5–7.5 cm long, 2–2.5 mm in diameter (excluding glume tips), axillary and terminal, pencil-shaped with divergent glumes, often partly hidden by sheaths. Sessile spikelets completely sunken in the rachis, the two glumes similar, oblong with acute-acuminate apices, the first glume covering the

FIG. 49. *Rhynchelytrum repens* (*Rogers 96-99*)

rest of the spikelet (as seen from outside the inflorescence), the second glume lining the cavity in the axis occupied by the spikelet, the lower lemma matching the glumes in shape and size, with well-developed palea; upper lemma shorter and thinner textured than the lower one, likewise with well-developed palea. Pedicellate spikelets on flat, strap-shaped pedicels; first glume lanceolate, flattened into a scale appressed against the inflorescence and covering the rest of the spikelet (this the glume often seen diverging from the inflorescences); second glume lanceolate,

Fig. 50. *Rottboellia cochinchinensis* (*Rogers 98-5*)
Fig. 51. *Rottboellia cochinchinensis* (*Rogers 98-5*)

narrower than and shorter than the first glume, and covered by it, the floret solitary, its lemma and palea delicate. Inflorescences breaking into small cylindric diaspores provided with basal elaiosomes.

Field recognition characters.—Very tall grass with long brittle hairs on sheaths; inflorescence a narrow cylinder.

Common name.—Corn Grass.

Distribution and habitat.—*Rottboellia cochinchinensis* is an introduced Old World species abundant in Barbados, especially in low moist settings. This tall grass is provided with stiff, brittle hairs that can become painfully embedded in the flesh.

Comment.—This species is often listed under the illegitimate name "*Rottboellia exaltata.*"

Representative specimens: **BARBADOS. Christ Church:** Meadow just N of ABC Highway near Kingsland, 11 Sep 1996, *Rogers 96-48* (BAR). **St. George:** Rowan's Plantation, 1 Oct 1958, *Hunte s.n.* (BAR); Rowan's, 15 Mar 1959, *McKenzie s.n.* (BAR). **St. James:** Holetown, along stream just E of the sea, 10 Oct 1996, *Rogers 96-78* (BAR). **St. Joseph:** Weed at Andromeda Botanic Gardens, 27 Jan 1998, *Rogers 98-5* (BAR). **St. Philip:** Junction of Highway 4B and road into Golden Grove, roadside clump, 23 Oct 1996, *Rogers 96-112* (BAR).

Schizachyrium Nees, Fl. Bras. Enum. Pl. 331. 1829.

Inflorescences terminal and axillary, each a single flowering stem (rame) on a peduncle. Spikelet pairs disarticulating with rachis segments; one spikelet of the pair sessile, awned on the upper lemma and perfect, the other pedicellate and reduced. Glumes nearly equal, both lemmas soft-textured.

A genus of some 50–60 species from warm regions worldwide. *Schizachyrium* is related to and sometimes included in *Andropogon,* and differs by having just one flowering stem (rame) per peduncle.

62. Schizachyrium brevifolium (Sw.) Nees ex Buse, Pl. Jungh. 359. 1854. BASIONYM: *Andropogon brevifolius* Sw., Prodr. 26. 1788.

Delicate grass. Culms to ca. 40 cm tall, mostly under 1.5 mm in diameter. Sheaths 18–30 mm long, auriculate and finely pubescent at the collar. Ligule 0.25 mm long, white (dry), ciliate. Blades up to 4.5 cm long, 1–6 mm wide, linear to ligulate or oblong, frequently with rounded apices. Inflorescences ca. 1 cm long (excluding awns) partly hidden by sheaths, with fewer than 10 spikelets showing, these provided with awns 5 mm long.

Field recognition characters.—Leaf blades often oblong with rounded apices. Inconspicuous, few-flowered inflorescences with long awns.

Habitat and distribution.—A pantropical species known from only one locality in Barbados and not encountered in recent collections. The distributional remark in Gooding et al. (1965), "confined to one or two gullies," is inconsistent with the only collection in the Barbados Herbarium, as cited below. Searching the dry fields at that locality during January 1998 failed to turn up this species. The area has some open meadows remaining among sugar fields and residential developments. Most of the grass cover at the site is *Bothriochloa bladhii.*

Comment.—Our description is based on the only Barbadian specimen.

Representative specimen: **BARBADOS. St. Peter:** Maynards, dry field, Jan 1939, *Allan 510* (BAR).

Setaria P. Beauv., Ess. Agrostogr. 51. 1812, nom. cons.

Inflorescence a more or less cylindric, tightly congested panicle bearing bristles among the spikelets, these not disarticulating with the bristles. Spikelets with 2 florets, the lower one reduced, the upper one perfect. First glume reduced. Second glume and lower lemma of similar size. Upper lemma firm with inrolled margins, tending to be transverse-rugose.

A genus of over 100 species of the Old and New Worlds.

KEY TO *SETARIA* SPECIES IN BARBADOS

1. Blades generally wider than 1 cm .... **Setaria barbata**
1. Blades narrower than 1 cm 2
   2. Spikelets uniformly crowded along panicle; inflorescence axis puberulent .... **Setaria parviflora**
   2. Spikelet density along main axis variable, with stretches of bare axis apparent; inflorescence axis hirsute .... **Setaria tenax**

63. Setaria barbata (Lam.) Kunth, Revis. Gramin. 1:47. 1829. (**Fig. 52**). BASIONYM: *Panicum barbatum* Lam., Tabl. Encycl. 1:171. 1791.

Broad-leaved, shade-loving grass. Culms rising to ca. 80 cm tall, sometimes partly sprawling, often with roots at multiple basal nodes, becoming woody toward the base on large specimens, the nodes dark. Sheaths 9–15 cm long, sometimes hirsute. Blades to 32 cm long and 30 mm wide, elliptic to lanceolate, with a conspicuous white midrib when dry, having a long-attenuate apex, hirsute with pustule-based hairs on both surfaces and with a scaberulous underlayer. Panicles 16 cm long, 5 cm wide, pyramidal, open, the branches to 3.5 cm long, uncrowded, straight and spreading to ascending, densely covered with spikelets; secondary branches (when present) very short, their axes obscured by the dense covering of spikelets on the primary branches. Inflorescence main axis hirsute with a puberulent-scabrous understory. Spikelets 2.2–3.0 mm long, subtended by 0–1(–2) sinuous, antrorse-barbed bristles to ca. 7 mm long. First glume 1/3 the length of the spikelet, ovate, acute or obtuse apically, 5-nerved. Second glume over half the length of the spikelet, lance-elliptic, 5–7-nerved, often mucronulate. Lower lemma as long as the spikelet, soft-textured, oblong, acute at the apex, 7-nerved, with palea. Upper lemma matching the lower one in size and shape, white-indurate, transversely rugose.

Field recognition characters.—Shade-loving grass with broad leaf blades; panicle with bristles subtending the spikelets.

Distribution and habitat.—An introduced Asian species, *Setaria barbata* is frequent in Barbados in shaded places.

Representative specimens: **BARBADOS. St. James:** Holetown, at edge of lawn under shadetree, 10 Oct 1996,

Rogers 96-79 (BAR, FLAS). **St. John:** Deep shade on grounds of Codrington College, 23 Oct 1996, *Rogers 96-113* (BAR); Newcastle, roadside, 22 Feb 1990, *Carrington & Taylor 1193* (BAR); Montcrieffe, under trees near estate house, Oct 1940, *Gooding* 7 (BAR); Montcrieffe, gully, talus slope, Nov 1940, *Gooding* 7 (BAR). [Note repeated "species" number.]

64. **Setaria parviflora** (Poir.) Kerguelen, Lejeunia n.s., 120:161. 1987. BASIONYM: *Cenchrus parviflorus* Poir., Encycl. 6:52. 1804.

Setaria geniculata P. Beauv., Ess. Agrostogr. 51, 169, 178. 1812.

Upright weedy grass. Culms straw-colored (dry) with black (dry), glabrous nodes. Sheaths up to 8.5 cm long, glabrous. Ligule 0.5–0.75 mm long, a fringe of cilia. Blades 7–12 cm long, 3–6 mm wide, widest below the middle and with long-attenuate apices, both surfaces scabrate. Cylindric panicles 3–5 cm long, up to 1.5 cm wide, tawny, with a hint of purple in the bristles (dry). Spikelet subtended by a cluster of numerous antrorse bristles up to ca. 6 mm long. Spikelets ca. 2.2–2.5 mm long. First glume about 1/3 the length of the spikelet, ovate-deltoid with rounded-acute apex, trinerved. Second glume similar but larger (a little over half the length of the spikelet), trinerved. Lower lemma as long as the spikelet, lanceolate, soft-textured, with a palea. Upper lemma indurate, transversely rugose.

Field recognition characters.—Narrow-leaved (blades to 6 mm wide) foxtail-bearing grass with the bristles not dehiscing with the spikelets, and with the spikelets uniformly crowded along the main axis.

Distribution and habitat.—*Setaria parviflora* ranges from the U.S. to South America. It is recorded from Barbados from one 1940 collection in a roadside ditch. Our description is based on that.

Representative specimen: **BARBADOS. St. Thomas:** Canefield, roadside ditch, Jul 1940, *Gooding* 6 (BAR).

65. **Setaria tenax** (Rich.) Desv., Opusc. Sci. Phys. Nat. 78. 1831. BASIONYM: *Panicum tenax* Rich., Actes Soc. Hist. Nat. Paris 1:106. 1792.

Upright grass. Culms sometimes exceeding 40 cm tall. Sheaths to 9 cm long, ciliate along the upper margins, and with a beard at the collar. Blades 15.5–23 cm long, 5–ca. 9 mm wide, serrulate-margined, adaxially hispidulous, abaxially glabrous. Panicle ca. 18 cm long, up to 20 mm wide (including bristles), narrowly cylindric, with the spikelets arranged in clumps along the main axis, this exposed in places, hirsute. Spikelets ca. 2.2–2.5 mm long, subtended by clusters of numerous, flexuous, antrorsely and retrorsely barbed bristles to 1.5 cm long, these with silky cilia basally. First glume 1/3 the length of the spikelet, broadly ovate, mucronulate. Second glume a little over half the length of the spikelet, broadly ovate. Lower lemma lanceolate, as long as the spikelet, delicate-textured. Upper lemma as long as the spikelet, lanceolate, white-indurate, transversely rugose.

Field recognition characters.—Blades long and narrow with a beard at the collar; panicle narrowly cylindric, pubescent on the axis, bristly; spikelets in clumps.

Distribution and habitat.—*Setaria tenax* occurs from Mexico to South America. It is rare in Barbados.

Representative specimens: **BARBADOS. St. Andrew:** Walker's Savannah, scrub, 3 Sep 1997, *Carrington 1787* (BAR, FLAS); Turner's Hall Woods, along partly shaded dirt road, past its flowering peak and mostly dried and dead, scarce, 27 Jan 1998, *Rogers 98-14* (BAR).

Sorghum Moench, Methodus 207. 1794, nom. cons.

Large, coarse, upright grasses. Panicles with spikelets in pairs or in triplets: one sessile, the other one or 2 spikelets pedicellate and thinner. Florets 2 per spikelet, with thin-textured lemmas.

An Old World, primarily African genus of about 50 species.

KEY TO *SORGHUM* SPECIES IN BARBADOS

1. Perennial with rhizome; leaf blades to 2 cm wide; panicle uncrowded, to 40 cm long; sessile spikelet sometimes awned; caryopsis remaining within glumes .... **Sorghum halepense**
1. Annual with no rhizome; leaf blades to 5 cm wide; panicle compact, to 20 cm long; sessile spikelet usually awnless; caryopsis often exposed outside of glumes .... **Sorghum bicolor**
   [Cultivated sorghum and "guinea corn," not seen in Barbados outside of cultivation.]

FIG. 52. *Setaria barbata* (*Rogers 96-113*)
FIG. 53. *Sorghum halepense* (*Rogers 96-131*)

66. **Sorghum halepense** (L.) Pers., Syn. Pl. 1:101. 1805. (**Fig. 53**). BASIONYM: *Holcus halepensis* L., Sp. Pl. 1047. 1753.

Large, coarse, upright weedy grass. Sheaths glabrous. Ligules 2 mm long, ciliate. Blades to 32 cm long and 2 cm wide, very narrowly lanceolate, with a conspicuous white midrib and 10–12 pronounced (abaxially) longitudinal nerves, tapered to a filiform apex, glabrous except for the scabrous margin. Panicles pyramidal, open and uncrowded, 30–43 cm long, 15–25 cm wide, the primary branches to 14.5 cm long, sub-branched; axes glabrate, scabrous or (sometimes distally) hirsute. Spikelets 5–6 mm long (excluding anthers, styles, and awns). Sessile spikelet far more plump than the pedicellate ones, the glumes covered densely with appressed ascending hairs; first glume as long as the spikelet, lanceolate, tridentate apically; second glume almost as long, somewhat indurate with a keeled midrib; lower lemma and palea linear-lanceolate, hyaline and very delicate, more or less as long as the spikelet, pubescent; upper lemma reduced, with 2 lanceolate lobes, armed between the lobes with a deciduous (often absent), twisted, hygroscopic awn 6–14 mm long. Pedicellate spikelets narrow and sharp-pointed; first glume as long as the spikelet and wrapped partly around it, 5-nerved, narrowly lanceolate, ascending-hirsute; second glume almost as long as the first glume but narrower, similarly pubescent, keeled; the two florets both smaller than the glumes, long and narrow, with very delicate lemmas and paleas.

Field recognition characters.—Coarse grass with large (generally over 30 cm long), open pyramidal panicle; spikelets 5–6 mm long, often awned from deep within (on a reduced lemma), the awn twisted, geniculate, and hygroscopic.

Common name.—Johnson Grass.

Distribution and habitat.—The present work is the first record for this introduced Old World grass in Barbados, which is surprising, given the abundance and conspicuousness of the species here. That it is not included in Gooding et al. (1965) suggests the present abundance of "Johnson Grass" to be relatively recent in Barbados. As the species occurs today, earlier collectors would not have overlooked it.

Representative specimens: **BARBADOS. Christ Church:** Sandy area between Graeme Hall Swamp and Worthing Beach, weed, 5 Jan 1997, *Rogers 97-4* (BAR). **St. Michael:** Roadside weed along Highway 4 just E of intersection with ABC Highway, 9–15 Nov 1996, *Rogers 96-131* (BAR, FLAS).

Sporobolus R. Br., Prodr. 169. 1810.

Panicles with one-flowered spikelets articulating above the glumes. First glume smaller than the second glume.

A genus of some 100 species found around the globe in warm climates.

KEY TO *SPOROBOLUS* SPECIES IN BARBADOS

1. Spikelets under 1 mm long, uncrowded .... **Sporobolus tenuissimus**
1. Spikelets longer than 1 mm, crowded 2
   2. Panicle branches devoid of spikelets at their bases .... **Sporobolus coromandelianus**
   2. Panicle branches bearing spikelets for their entire lengths 3
      3. Panicle usually longer than 12 cm; leaf blades often longer than 12 cm; plants not rhizomatous; spikelets shorter than 2 mm .... **Sporobolus indicus** var. **pyramidalis**
      3. Panicle usually shorter than 12 cm; leaf blades shorter than 12 cm; plants rhizomatous; spikelets longer than 2 mm .... **Sporobolus virginicus**

67. Sporobolus coromandelianus (Retz.) Kunth, Revis. Gramin. 1:68. 1829. (**Fig. 54**). BASIONYM: *Agrostis coromandelianus* Retz., Observ. Bot. 4:19. 1786.

Sporobolus pyramidatus (Lam.) Hitchc., Man. Grasses W. Ind. 84. 1936. BASIONYM: *Agrostis pyramidata* Lam., Tabl. Encycl. 1:161. 1791.

Grasses of variable habit. Culms to 40 cm tall, sprawling or erect. Sheaths mostly glabrous though with some marginal cilia, these most abundant at the collar, here the hairs often forming a long tuft. Ligule ca. 1 mm long, a band of white hairs. Blades to 12 cm long and 2.5 mm wide, keeled, glabrous with scabrous margins. Inflorescences 9 cm long and 3 cm wide, pyramidal when fully open (often encountered with the branches appressed-ascending), light tawny to nearly white, the major branches generally unbranched, devoid of spikelets basally. Inflorescence axes glabrous or minutely scabrate. Spikelets usually ca. 1.5–1.7 mm long, short-pedicellate or subsessile along sinuous axes. First glume ca. 1/4 the length of the spikelet, oblong, rounded or obtuse apically. Second glume nearly as long as the spikelet, oblong, acute at the apex, uninerved and keeled. Lemma the longest organ of the spikelet, oblong, apically acute, uninerved, its palea almost as long.

Field recognition characters.—Coastal grasses having pyramidal inflorescences with primary branches not floriferous near base.

Distribution and habitat.—*Sporobolus coromandelianus* is an attractive salt-loving species encountered from the U.S. to South America.

Representative specimens: **BARBADOS. St. Andrew:** On hills near Belleplaine, overlooking the Coastal Highway, adjacent to seashore, 11 Sep 1996, *Rogers 96-43* (BAR, FLAS). **St. Lucy:** River Bay, on rocks overlooking the sea, 19 Dec 1996, *Rogers 96-163* (BAR). **St. Michael:** Bridgetown, Careenage, 14 Nov 1996, *Forde 96-20* (BAR, FLAS). **St. Philip:** Bel Air, dry rocky pasture, Jul 1940, *Gooding 519* (BAR); East Point Lighthouse, rocky area near cliffs, 7 Nov 1996, *Forde 96-8* (BAR).

68. Sporobolus indicus var. **pyramidalis** (P. Beauv.) Veldk., Blumea 35:439. 1991. (**Figs. 56, 57**). BASIONYM: *Sporobolus pyramidalis* P. Beauv., Fl. Oware 2:36. 1816.

Sporobolus jacquemontii Kunth, Revis. Gramin. 2:427, pl. 127. 1831.

Tufted, upright, slender grass. Culms to ca. 1.2 m tall with dark nodes. Sheaths to 20(–26) cm long. Blades to 50 cm long and 1–2(–6) mm wide, mostly basal, with some inserted at acute angles on upper nodes, linear. Panicles 13–40 cm long, usually narrower than 2 cm (when branches not spread), vertical, usually tightly congested and narrowly cylindric-fusiform with the primary branches appressed-ascending (or sometimes spreading with age), the primary branches to 5 cm long. Spikelets 1.6–1.8(–2.1) mm long, pedicellate, glabrous or scaberulate on

FIG. 54. *Sporobolus coromandelianus (Forde 96-5)*
FIG. 55. *Sporobolus tenuissimus (Rogers 97-68)*

the glumes. First glume ca. 1/4–1/3 the length of the spikelet, oblong, with a truncate-erose apical margin. Second glume 1/3–1/2(or more) the length of the spikelet, longer than the first glume, elliptic or oblong, or sometimes obovate, one-nerved, truncate (or rounded or acute) apically, with erose irregularities in the apical margin. Lemma and palea defining the length of the spikelet, the lemma narrowly lanceolate, uninerved, the palea as long as the lemma, elliptic, slightly tricuspidate apically. Grain falling readily from the foliar-floral organs, rugose.

Field recognition characters.—Tufted upright grasses with long wire-thin leaves and usually with wandlike (most often unexpanded) vertical, tightly congested panicles.

Common name.—Wire Grass.

Distribution and habitat.—Interpreted broadly, this species is pantropical. In Barbados it is a common weed, often found near the sea.

Comments.—Determination of the circumscription and appropriate name for grasses variously called *Sporobolus jacquemontii, S. poiretii, S. indicus, S. pyramidalis* P. Beauv. [not to be confused with "*S. pyramidatus* (Lam.) Hitchc."], *S. tenuissimus*, and others have been areas of disagreement and confusion rooted in the highly polymorphic pantropical "*Sporobolus indicus* complex" (see e.g., Clayton, 1964; Baaijens & Veldkamp, 1991; Veldkamp, 1990). The name *Sporobolus poiretii* used in Gooding et al. (1965) is, according to Baaijens & Veldkamp (1991, p. 401), the result of "enormous confusion" caused by Hitchcock's (1936, 1950) misapplication over several years of the name "*S. poiretii*" to *Sporobolus indicus.* Baaijens & Veldkamp (1991) interpreted *Sporobolus indicus* to embrace much of the complex, including *S. jacquemontii.* They placed both *S. jacquemontii* and *S. pyramidalis* in synonymy under *S. indicus* var. *pyramidalis*. We find the key characters used to distinguish var. *pyramidalis* (erosely truncate upper glumes slightly longer than lower glume and less than half as long as the spikelet) to identify most Barbados material. Some Barbados specimens deviate, however, by having the upper glume with an acute apex, and some spikelets

FIG. 56. *Sporobolus indicus* var. *pyramidalis* (*Rogers 96-124*)
FIG. 57. *Sporobolus indicus* var. *pyramidalis* (*Rogers 96-124*)

have the second glume longer than half the length of the spikelet. Davidse & Pohl (1994) regard *S. jacquemontii* as a separate species, and our material would fall under their *S. jacquemontii.* We tentatively prefer Baaijens & Veldkamp's (1991) conservative approach of a broadly conceived *Sporobolus indicus.*

Representative specimens: **BARBADOS. Christ Church:** Maxwell Coast, sandy soil, May 1940, *Gooding 10* (BAR); Atlantic Shores, weed on thin soil in pockets on rocks overlooking the sea, 24 Aug 1996, *Rogers 96-24* (BAR, FLAS). **St. Andrew:** Scotland District, wet soil on stream bank near Haggats, 21 Aug 1996, *Rogers 96-16* (BAR). **St. Joseph:** Highway 3A, hilltop meadow near Sturges, 6 Sep 1996, *Rogers 96-42* (BAR); roadside between Sturges and Castle Grant in low wet area, 4 Oct 1996, *Rogers 96-80* (BAR, FLAS). **St. Peter:** Speightstown, beach front, 29 Oct 1996, *Rogers 96-124* (BAR, FLAS).

69. **Sporobolus tenuissimus** (Mart. ex Schrank) Kuntze, Revis. Gen. Pl. 3:369. 1898. (**Fig. 55**).

BASIONYM: *Panicum tenuissimum* Mart. ex Schrank, Denkschr. Konigl.-Baier. Bot. Ges. Regensburg 2:26. 1822.

Delicate, tufted, upright, slender, glabrous, weedy grass. Culms up to ca. 1 m tall. Sheaths up to ca. 9 cm long, laterally compressed. Ligules ca. 0.1 mm long, erose membranes. Blades 10–33 cm long, 1–3(–4) mm wide, acicular, widest at the base. Panicles to a little over 58 cm long and 12 cm wide, narrowly pyramidal to fusiform, the branches threadlike, the primary branches up to ca. 7 cm long, numerous, straight and rebranched. Spikelets under 1 mm long, pedicellate. First glume much shorter than the second glume, truncate-rounded and erose apically. Second glume half as long as the lemma, lance-elliptic with an acuminate apex. Lemma lanceolate, its palea as long. Achene widest distally.

Field recognition characters.—Weedy grasses with long, open panicles having threadlike axes and the tiny (under 1 mm long) spikelets borne singly or nearly so on long pedicels.

Distribution and habitat.—Distributed from the West Indies to Brazil, *Sporobolus tenuissimus* is an abundant weed seasonally.

Comments.—The tall but very delicate panicle is distinctive, suggesting a large *Eragrostis* at first glance (*Eragrostis* differs by having multiple florets per spikelet).

Representative specimens: **BARBADOS. St. Joseph:** Horse Hill, disturbed weedy ground behind the St. Joseph Post Office, 19 Nov 1997, *Rogers 97-54* (BAR, FLAS); Andromeda Botanic Gardens, weed in propagation area, 22 Dec 1997, *Rogers 97-68* (BAR).

70. Sporobolus virginicus (L.) Kunth, Revis. Gramin. 67. 1829. (**Figs. 58, 59**). BASIONYM: *Agrostis virginica* L., Sp. Pl. 63. 1753.

Variable grass, either prostrate and creeping over (and under) beach sand or short-tufted and erect. Culms to (rarely) ca. 1 m tall. Leaves often inserted rung-like and spreading along the culm. Sheaths 1–7 cm long, glabrous except for occasional apical-marginal silky cilia. Ligules ca. 0.1 mm long, a fringe of cilia. Blades 4–12 cm long and up to 5 mm wide at the base (the widest point of the blade), acicular, adaxially ridged, the ridges provided with dimorphic papillae: one type short and rounded, the other type far longer, stout, and sharp-pointed, these also comprising the minute teeth along the leaf margin, with a long silky tuft at the collar, this sometimes extending onto the adaxial surface of the blade. Inflorescences (2–)4.5–9 cm long, 0.3–1 cm wide, congested panicles with appressed upright branches, usually straw-colored to nearly white (fresh and dry), held well above the foliage. Spikelets 2.4–2.8 mm long, glabrous. First glume 2/3–3/4 the length of the lemma, lance-elliptic, uninerved. Second glume almost as long as the lemma, lance-elliptic, mucronulate, uninerved. Lemma resembling the second glume but a little longer. Palea similar to the lemma and almost as large.

Field recognition characters.—Often encountered as a strongly rhizomatous, sand-binding species; leaves short and acicular, strongly two-ranked; inflorescence usually nearly white, erect, vertical, fusiform. Vegetatively similar to *Paspalum vaginatum*.

Common name.—Crab Grass.

Distribution and habitat.—*Sporobolus virginicus* is one of the predominant sand-binding beach and dune grasses in Barbados. It is also found on rocky limestone- and sandstone-dominated hills rising above the East Coast beaches, and in brackish Graeme Hall Swamp. *Sporobolus virginicus* ranges from the U.S. to South America.

Comments.—Although very different in the structure of the inflorescence and the spikelets, *Sporobolus virginicus* is bewilderingly similar in vegetative aspect and in habitat to *Paspalum vaginatum.* See discussion of *P. vaginatum.*

Representative specimens: **BARBADOS. Christ Church:** Swamp, Chancery Lane, 24 June 1990, *Carrington & Taylor 1234* (BAR); seashore, Maxwell's Coast, Jan 1940, *Gooding 467* (BAR). **St. Andrew:** Sandy dune area between beach and E Coast Rd. between Belleplaine and Barclays Park, 21 Aug 1996, *Rogers 96-21* (BAR); beach S of Belleplaine and N of Barclays Park, 18 Oct 1996, *Rogers 96-92* (BAR, FLAS). **St. Peter:** Mullins Beach, on the beach, 29 Oct 1996, *Rogers 96-119* (BAR); Speightstown, 29 Oct 1996, *Rogers 96-122* (BAR).

Stenotaphrum Trin., Fund. Agrost. 175. 1820.

Prostrate grasses rooting at the nodes. Inflorescences with a thickened axis having the branches embedded in it. Spikelets disarticulating with rachis segments. Spikelets 2-flowered, the upper floret perfect, the lower one staminate or neuter.

A small tropical and subtropical genus.

71. Stenotaphrum secundatum (Walter) Kuntze, Revis. Gen. Pl. 2:794. 1891. (**Figs. 60, 61**). BASIONYM: *Ischaemum secundatum* Walter, Fl. Carol. 249. 1788.

Culms prostrate and rooting at the nodes. Leaves in decussate clusters often separated by long bare internodes. Sheaths 1.5–4.5(–7) cm long, laterally compressed and conduplicate, often pulling away from the stem, glabrous except for an apical-marginal band of fringe. Ligule ca. 0.5–0.75 mm long, a uniform, light-colored band of cilia. Blades 1.9–12.5 cm long, 4–8 mm wide (when flattened), conduplicate, rounded or truncate apically, with a small notch at the collar. Inflorescences 2.8–6.5 cm long, 3–6 mm wide, formed into a flattened, fleshy main axis. Spikelets 4–5 mm long, appressed to and partly embedded in the fleshy axis. First glume ca. 1/3 the length of the spikelet, ovate with an acute apex. Second glume as long as the spikelet, appressed to the inflorescence axis, lanceolate with an acuminate apex, 7-nerved. Lower lemma

FIG. 58. *Sporobolus virginicus* (*Rogers 96-122*)
FIG. 59. *Sporobolus virginicus* (*Rogers 96-122*)

almost as long as the spikelet, 7-nerved, its palea much smaller. Upper lemma lanceolate, a little shorter than the lower one and softer in texture. Upper palea almost identical to the upper lemma.

Field recognition characters.—Prostrate grass with blunt folded leaf blades; conduplicately flattened, nested sheaths; and thickened, flattened inflorescence axes.

Common name.—St. Augustine Grass.

Distribution and habitat.—*Stenotaphrum secundatum* is pantropical and is found in subtropical regions; it is used as a coarse turf grass. This species grows uncultivated in Barbados especially on seaside dunes and rocky hillsides overlooking the sea.

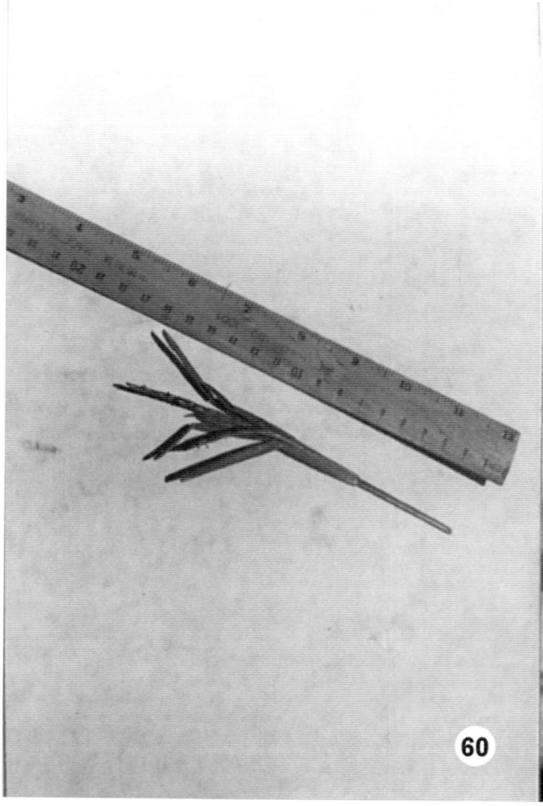

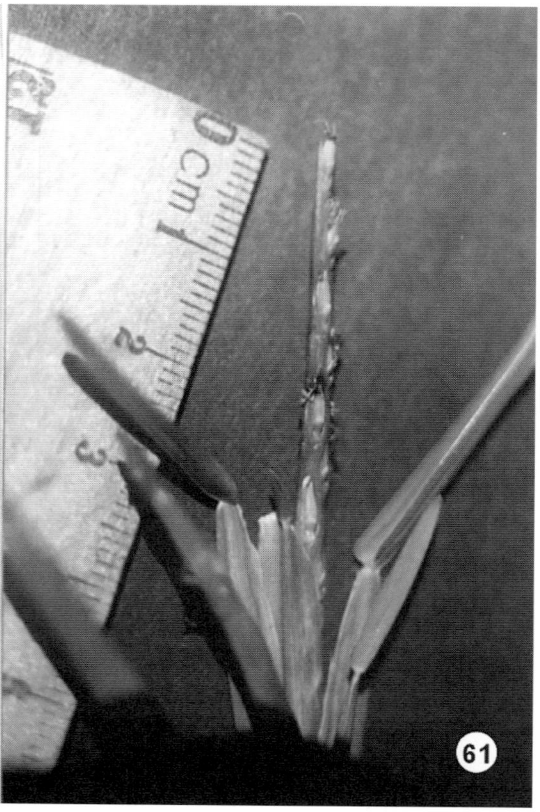

FIG. 60. *Stenotaphrum secundatum* (*Rogers 96-94*)
FIG. 61. *Stenotaphrum secundatum* (*Rogers 96-94*)

Representative specimens: **BARBADOS. St. Andrew:** On dunes behind beach, between Belleplaine and Barclays Park, 18 Oct 1996, *Rogers 96-94* (BAR, FLAS). **St. John:** Lodge School, cricket field in shelter of wall, Jul 1941, *Gooding 511* (BAR). **St. Joseph:** Bathsheba, seaside grassland, Feb 1941, *Gooding 511* (BAR—note repeated species number).

Themeda Forssk., Fl. Aegypt.-Arab. 178. 1775.

Inflorescences with spikelets tightly congested in fan-shaped fascicles, these made up of long narrowly lanceolate glumelike foliar organs, sessile spikelets, and pedicellate spikelets. Some spikelets with extremely long awns.

A small Old World genus.

KEY TO *THEMEDA* SPECIES IN BARBADOS

1. Awns mostly longer than 6 cm; hair bases on sterile glumes not swollen, or swollen and conical to cylindric, seldom globular .... **Themeda arguens**
1. Awns mostly shorter than 5 cm; hair bases on sterile glumes dramatically swollen and resembling droplets of liquid even when dry .... **Themeda quadrivalvis**

These two species are confusingly similar, and authors differ on the characters used to distinguish them. Gooding et al. (1965) felt both to be present, and we concur, although the characters used by those authors fail to give a clear separation. We have made use of an herbarium loan of both species from the Royal Botanic Gardens Kew (K) to sharpen our species concepts. Both species were collected during different seasons on Horse Hill, St. Joseph Parish. The collection *Rogers 98-15* from near Horse Hill appears intermediate.

72. Themeda arguens (L.) Hack., Monogr. Phan. 6:657. 1889. BASIONYM. *Stipa arguens* L., Sp. Pl. Ed. 2. 117. 1762.

Culms to ca. 2 m tall. Sheaths to 7 cm long, glabrous to villous. Ligules ca. 1 mm long, membranous. Blades to 24 cm long and 7 mm wide, adaxially smooth to scabrous, abaxially smooth or nearly so, the margins scabrous, otherwise glabrous except for occasional pustule-based adaxial hairs around the collar. Inflorescences axillary, flabellate, the bracts to 9(–18) cm long and 3 mm wide. Units of the inflorescence tending toward larger dimensions than in *T. quadrivalvis*: glumes to ca. 4–5 cm long, the awns (4.5–)6–7(–9.5) cm long. Hairs on the glumes without swollen bases or with variable swollen bases, these generally pyramidal or cylindric, or occasionally globular.

Field recognition characters.—Inflorescences flabellate with awns mostly over 6 cm long.

Common name.—Piano Grass.

Distribution and habitat.—An Asian species introduced as fodder. Uncommon in Barbados.

Representative specimens: **BARBADOS.** No locality data. 1927, *Ballou s.n.* (US-photocopy examined). **St. Joseph:** Pasture by Grantley Adams School, 28 Jan 1998, *Rogers 98-15* (BAR); Road cut at top of Horse Hill, 19 Nov 1997, *Rogers 97-57* (BAR, FLAS). **St. Michael:** Codrington Botanic Station, Nov 1935, *Mayers 8* (BAR, K). **St. Thomas:** Hilltop meadow between Flower Forest and Sturges, covering a large area as a monospecific stand, 16 Dec 1997, *Rogers 97-63* (BAR, FLAS).

73. **Themeda quadrivalvis** (L.) Kuntze, Revis. Gen. Pl. 2:794. 1891. BASIONYM: *Andropogon quadrivalvis* L., Syst. Veg. ed. 13. 758. 1774.

Culms over 1 m tall, arching. Sheaths to ca. 6 cm long, mostly glabrous. Ligules ca. 1.2 mm long, membranous flaps with irregular margins. Blades to 17 cm long, 6 mm wide, scabrous-margined and scabrous adaxially, otherwise glabrous or nearly so. Inflorescences tightly congested, fan-shaped fascicles on short, wiry peduncles in leaf margins, having narrowly lanceolate, hirsute bracts and bracteoles usually under 6 cm long. Apart from bracts and bracteoles, the inflorescences consisting of 3 basic units: 1) An outer cluster of sessile, narrowly lanceolate glumelike foliar organs usually ca. 3–5 cm long, these hirsute with conspicuously pustule-based hairs, and with long-attenuate acicular apices. 2) Pedicellate awnless spikelets, these with the larger glume resembling the foliar organs described above, the smaller glume ca. 1 cm long, with a somewhat shorter empty lemma and with 2 additional chartaceous, sterile, lemmalike organs (possibly a sterile lemma and palea), the smaller glume, sterile lemma, and lemmalike organs resembling each other. 3) Awned spikelets having the larger glume to ca. 3 cm long, narrowly lanceolate, hirsute with swollen-based hairs, and with a long-attenuate acicular tip; shorter glume ca. 1–1.5 cm long, subulate; sterile lemma ca. 1 cm long and similar to the shorter glume; fertile lemma variously pubescent to nearly glabrous, ca. 8 mm long, indurate and wrapped around the palea, nested in long, stiff bristles, surrounding an awn 3.6–5 cm long, and with an additional small, linear, apparently sterile floret inserted above the fertile one.

Field recognition characters.—Fan-shaped inflorescences with glumes ca. 3 cm long and bearing awns usually 4–5 cm long.

Distribution and habitat.—An introduced Old World species, occasionally encountered in Barbadian meadows, especially near the upland center of the island.

Representative specimens: **BARBADOS. St. Joseph:** Along Highway 3 at base of rocky highway cut on Horse Hill, 14 Jan 1997, *Rogers 97-12* (BAR, FLAS). **St. Michael:** Garrison Savannah, "3/5/27," *Comm. C. C. Skeete s.n.* (K, US-photocopy examined).

Urochloa P. Beauv., Ess. Agrostrogr. 52. 1812.

Inflorescence with uncrowded straight side-branches bearing the spikelets on one side, the secondary branching tending to be minimal (in ours). First glume short, tending to be oriented toward the rachis. Second glume and lower lemma of similar length. Upper lemma cartilaginous, rugose, often apiculate, with the margins enfolding the palea.

A controversially circumscribed (sub)tropical genus of perhaps some 40 species. Authors differ concerning the morphological characterizations and circumscriptions of *Urochloa, Brachiaria*, and *Panicum*. The species included here in *Urochloa* have until recently usually been placed in *Brachiaria.* As discussed briefly by Davidese & Pohl (1994), this is an unresolved matter, and we have conformed almost entirely with them and with Webster & Peterson (1996) due to the breadth and recentness of these two related treatments. Webster & Peterson (1996) include in *Urochloa* the species generally known as (and referred to in the present work as) *Panicum maxi-*

mum. This species does have the rugose upper lemma that helps distinguish *Urochloa* from *Panicum*. *Brachiaria eruciformis*, the type species of *Brachiaria*, is included in the key to *Urochloa* because of its overall similarities to these species.

KEY TO *UROCHLOA* (AND *BRACHIARIA*) SPECIES IN BARBADOS

1. Grass mostly prostrate; leaf blades under 4 cm long; upper lemma mucronulate; bases of blades auriculate .... **Urochloa reptans**
1. Grass upright to sprawling; leaf blades mostly over 4 cm long; upper lemma without a mucro; bases of blades only rarely auriculate (in *Urochloa fasciculata*) 2
  2. Upper lemmas rugose with transverse ridges .... **Urochloa fasciculata**
  2. Upper lemmas smooth (all ridges longitudinal) 3
    3. Spikelets densely pubescent .... **Brachiaria eruciformis**
    3. Spikelets glabrous or nearly so 4
      4. Sheaths glabrous (except for marginal hairs); inflorescence with 3–4 primary branches .... **Urochloa subquadripara**
      4. Sheaths hirsute; inflorescence with more than 4 side branches .... **Urochloa mutica**

74. **Urochloa fasciculata** (Sw.) R.D. Webster, Austral. Paniceae 235. 1987 [1988]. BASIONYM: *Panicum fasciculatum* Sw., Prodr. 22. 1788. *Brachiaria fasciculata* (Sw.) Parodi, Darwiniana 15:96. 1969.

Culm rising to 90 cm tall, becoming woody, often much-branched basally with tufted ascending branches. Sheaths 5.5–10.5 cm long, hirsute toward the collar, and occasionally hispid-hirsute toward the base or on the sheath surface, then the hairs tending to be pustule-based. Blades 4–26.5 cm long, 3–14 mm wide, variable in size even on the same plant, often very narrowly lanceolate with auriculate bases, glabrous or nearly so except for scaberulous margins and sometimes with sparse trichomes on the abaxial surface, the broader blades with 10–12 conspicuous longitudinal nerves. Upswept panicle 2.5–10 cm long, 0.5–3.5 cm wide with unbranched or minimally branched side branches. Spikelets (2–)2.2–2.6 mm long, often becoming purplish (fresh and dry), pedicellate. First glume variable, shorter than half the length of the spikelet, ovate with the apex truncate to acute (or acuminate-mucronate), wrapped around the spikelet. Second glume as long as the spikelet, lance-elliptic and wrapped around the lemmas, 7-nerved, short-acuminate apically. Lower lemma as long as the spikelet, soft-textured, markedly rugose with transverse ridges. Upper lemma indurate, bony-white, rugose.

Field recognition characters.—Weedy tufted grass with culms and primary inflorescence branches upswept; blades often a little auriculate; spikelets often purplish, with the lower lemma rugose.

Distribution and habitat.—*Urochloa fasciculata* is a common weed of sunny meadows and roadsides. Its distribution extends from the U.S. to the Caribbean and South America.

Representative specimens: **BARBADOS. Christ Church:** Immediately E of Oistins on sunny, highly disturbed site along highway at turnoff to Miami Beach, 16 Aug 1996, *Rogers 96-6* (BAR, FLAS); in weedy field between Pearl Ave. and the South Point Lighthouse, at base of bluff, 4 Sep 1996, *Rogers 96-33* (BAR); on rocks behind beach, between Silver Sands Hotel and Silver Rock Hotel, 27 Sep 1996, *Rogers 96-74* (BAR, FLAS). **St. Andrew:** Turner's Hall Woods, open meadow, 19 Jan 1997, *Rogers 97-5* (BAR). **St. John:** Waste land near Codrington College, 21 Oct 1937, *McIntosh 445* (BAR). **St. Michael:** Elizabeth Dr., Pine Gardens, 3 Aug 1990, *Carrington & Howard 1323* (BAR).

75. **Urochloa mutica** (Forssk.) T. Q. Nguyen, Novosti Sist. Vyss. Rast. 1966:13. 1966. BASIONYM: *Panicum muticum* Forssk., Fl. Aegypt.-Arab. 20. 1775. *Brachiaria mutica* (Forssk.) Stapf, Fl. Trop. Afr. 9:526. 1919.

Brachiaria purpurascens (Raddi) Henr., Blumea 3:434. 1940. *Panicum purpurascens* Raddi, Agrostogr. Bras. 47. 1823.

Large aggressive, pubescent grass. Culms 3–7 mm in diameter, rising to ca. 1 m tall, or sprawling and rooting at the nodes, these bearded. Sheaths to 10(–15) cm long, conspicuously hirsute, especially so near the base of the blade. Blades 7–20(–23) cm long and 5–11(–15) mm wide, linear-lanceolate, long-attenuate, hirsute to glabrate on both sides, particularly pubescent marginally. Inflorescences up to 17(–19) cm long, 7–9 cm wide, the side branches 1.5–8(–11) cm long, shallowly sinuous, uncrowded, not (or minimally) branched, ascending. Spikelets 2.5–3 mm long, in unequally short-pedicellate pairs. First glume reduced, 1/3 the length of the spikelet. Second glume large and about the size of the lemma of the lower floret, this with a well-developed palea. Upper (perfect) floret with indurate lemma and with well-developed palea.

Field recognition characters.—Large stature with heavy pubescence, particularly hirsute sheaths and bearded nodes; inflorescences with long (1.5–8 or more cm), unbranched (or little-branched) side branches.

Distribution and habitat.—*Urochloa mutica* is an aggressive weed capable of taking over large areas, especially on moist meadows and shores of streams and ponds. It is invading a large portion of Graeme Hall Swamp on the South Coast of Barbados. This species is probably African (according to Pohl, 1980), and is widespread in the Neotropics.

Representative specimens: **BARBADOS.** No locality specified. 1900, *Bovell 247* (BAR). *Botanic Station Herbarium 247* (BAR). **Christ Church:** Graeme Hall Swamp, abundant, 5 Jan 1997, *Rogers 97-3* (BAR); Graeme Hall Swamp, damp muddy area near drainage ditch, "80/02/25," *Riven-Ramsey 57* (BAR); Graeme Hall Swamp, near the Coast Highway at edge of swamp, disturbed site, on hummock, 9 Nov 1996, *Rogers 96-130* (BAR, FLAS); Graeme Hall Swamp, full sun, moist trailside, 11 Jun 1997, *Rogers 97-24* (BAR); Atlantic Shores, overgrown meadow, 15 Oct 1996, *Rogers 96-81* (BAR). **St. Joseph:** Andrews Plantation, around pond, Dec 1935, *Mayers 315* (BAR); Sugar Hill, common weed between cultivated fields, 29 Nov 1996, *Rogers 96-139* (BAR). **St. Michael:** Bridgetown, wetland, N side Bridge Rd., Dec 1936, *Allan 315a* (BAR); Codrington House, "19.9.1904," *Reece 247* (BAR).

76. **Urochloa reptans** (L.) Stapf, Fl. Trop. Afr. 9:601. 1920. **(Figs. 62, 63).** BASIONYM: *Panicum reptans* L., Syst. Nat. ed. 10, 2:870. 1759. *Brachiaria reptans* (L.) C.A. Gardner & C.E. Hubb., Hooker's Icon. Pl. 34:3363:3. 1938.

Culms sometimes rising to ca. 35 cm tall, or prostrate, stoloniferous, and rooting at the nodes. Ligules ca. 1 mm long, bands of white hairs. Blades to 4 cm long and 9 mm wide, often lanceolate and basally auriculate, with scabrous margins and marginal silky hairs (mostly near the base), occasionally pubescent on the surfaces of the blades. Inflorescences with the side branches to 1.8 cm long. Spikelets 1.8 mm long, crowded, short-pedicellate, unequally paired. First glume 1/5 the length of the spikelet. Second glume and lower lemma similar in length and appearance. Upper floret rugose on the lemma, this mucronulate.

Field recognition characters.—Low and often prostrate grass with auriculate, more or less lanceolate leaf blades under 4 cm long.

Distribution and habitat.—*Urochloa reptans* is an Old World species sometimes forming mats on beach sand.

Representative specimens: **BARBADOS. St. Andrew:** Along coast between Belleplaine and Barclays Park, 18 Sep 1996, *Rogers 96-58* (BAR, FLAS). **St. George:** Gun Hill, in a grass ditch, Dec 1939, *Allan 503a* (BAR). **St. Lucy:** Animal Flower Cave, seashore, Jan 1938, *Allan 503b* (BAR). **St. Philip:** Ruby, 18 Aug 1904, *Bovell 538* (BAR).

77. **Urochloa subquadripara** (Trin.) R. Webster, Austral. Paniceae 252. 1987. BASIONYM: *Panicum subquadriparum Trin.*, Gram. Panic. 145. 1826. *Brachiaria subquadripara* (Trin.) Hitchc., Lingnan Sci. J. 7:214. 1929 [1931].

Culms rising to ca. 50 cm tall, or stoloniferous, then rooting at the nodes and mat-forming. Sheaths mostly glabrous except for long marginal hairs. Ligules ca. 1 mm long, brushlike. Blades up to 12.5 cm long, 0.7–1 cm wide, often crisped, the margins scabrous, ciliate toward the base; adaxially and abaxially glabrous to sparsely ciliate. Inflorescences with 3 or 4 racemose spikes up to 7.5 cm long. Spikelets subsessile, loosely arranged with the insertions 2.5 mm apart on one side of the sinuous, broadened rachis. First glume ca. 1/3 the length of the spikelet, broadly ovate. Second glume approximately the length of the spikelet. Sterile lemma resembling the second glume.

Field recognition characters.—Inflorescences with 3–4 straight, armlike side branches, these wide-spreading.

Distribution and habitat.—*Urochloa subquadripara* has not been reported previously for Barbados. It is a pasture grass indigenous to the Old World tropics.

Representative specimen: **BARBADOS. St. Andrew:** Along E Coast S of Belleplaine, on border between upper dunes and rocky hills, 18 Sep 1996, *Rogers 96-57* (BAR).

Vetiveria Bory, Bull. Sci. Soc. Philom. Paris 1822:43. 1822.

A genus of just 1(–10) species.

78. **Vetiveria zizanioides** (L.) Nash, Fl. S.E. U.S. 67, 1326. 1903. **(Figs. 64, 65).** BASIONYM: *Phalaris zizanioides* L., Mant. Pl. 2:183. 1771.

FIG. 62. *Urochloa reptans* (Silver Sands Park, Christ Church, Barbados)
FIG. 63. *Urochloa reptans* (*Rogers 96-58*)

Culms sometimes exceeding 2 m tall, up to 8 mm in diameter, tufted, upright. Sheaths to 26 cm long. Blades up to ca. 1 m long and ca.1 cm wide, linear, folded and keeled. Panicles 14–32 cm long, to ca. 7 cm broad, pyramidal or narrowly pyramidal, often purplish (fresh or dry), the wire-thin branches to ca. 8 cm long, whorled, ascending, usually bare toward the base. Spikelets ca. 4 mm long, paired, with one member of each pair sessile and the other pedicellate, armed on the glumes with stout, sharp, pustule-based hairs, 2-flowered. Sessile spikelets with the glumes of nearly equal size and as long as the spikelet; first glume with involute margins, lanceolate with an acuminate apex, showing just one prominent nerve; second glume lanceolate, trinerved; lower lemma as long as the glumes, vaguely trinerved, lanceolate, very delicate, almost transparent when moist, with abaxial silky cilia; upper lemma about as long as the glumes, extremely delicate, lanceolate with long-acuminate apex, uninerved, its palea much shorter, ligulate, delicate. Pedicellate spikelets similar, a little smaller, the glumes armed with pustule-based hairs; first glume lanceolate, uninerved; second glume similar, with marginal cilia; lower lemma as long as the glumes, lanceolate, 2-keeled, with marginal cilia; upper lemma as long as the glumes, very delicate, uninerved, its palea much smaller.

Field recognition characters.—Tall grasses with all foliage and axes ascending; blades up to ca. 1 m long, sharply folded; inflorescences large panicles with heavily armed spikelets.

Common names.—Khus Khus, Vetiver Grass.

Distribution and habitat.—Vetiver Grass is an Asian species used as a source of oils, and is popular for hedgerows and for erosion control. It frequently lines rural roads throughout Barbados.

Representative specimens: **BARBADOS. Christ Church:** Weed along road connecting Newton with Lower Greys, abundant, 12 Nov 1997, *Rogers 97-52* (BAR). **St. Andrew:** Weed in meadow, Turner's Hall Woods, 27 Jan 1998, *Rogers 98-9* (BAR). **St. John:** Along highway 4B at base of Mt. Pleasant, 23 Oct 1996, *Rogers 96-114* (BAR). **St. Joseph:** Adjacent to (just W of) Bathsheba in grazed area, in clumps 2 m or more tall, abundant, full sun, 14 Aug 1996, *Rogers 96-5* (BAR).

Fig. 64. *Vetiveria zizanioides* (*Rogers 96-114*)
Fig. 65. *Vetiveria zizanioides* (*Rogers 98-9*)

Zoysia Willd., Ges. Naturf. Freunde Berlin Neue Schriften 3:440. 1801, nom cons.

79. **Zoysia matrella** (L.) Merr., Phillip. J. Sci. 7:230. 1912. (**Figs. 66, 67**). Basionym: *Agrostis matrella* L., Mant. Pl. 2:185. 1771.

Zoysia tenuifolia Willd. ex Thiele, Linnaea 9(3):309. 1835.

Low, rhizomatous, carpet-forming grasses, often forming a humped mat. Culms sometimes rising to ca. 23 cm, glabrous except for occasional puberulence at nodes. Sheaths usually 1–2 cm long, glabrous except for cilia of varying lengths at the collar. Ligule a minute band of cilia. Blades straight, 1.5–8(–9) cm long, up to ca 2 mm wide when flattened, flat or involute. Inflorescence 15–20 mm long, vertically spicate, linear-fusiform. Spikelets ca. 3–3.5 mm long, pedicellate, appressed; with only one glumelike organ, this a folded envelope around the single floret; the floret with just one foliar organ, this a folded envelope smaller than the "glume" but otherwise resembling it and enclosing the floret.

Field recognition characters.—Blades needle-thin, usually involute; inflorescence spicate and under 2 cm long.

Common name.—Zoysia.

Distribution and habitat.—An introduced Old World species escaped in Barbados.

Comments.—*Zoysia matrella* is an Asian grass popular for lawns, especially near the sea for its salt tolerance. This species thrives as an escape along Barbados shores, often with direct exposure to salt spray or to breaking waves. Our species concept is based on Davidse & Pohl (1994).

Representative specimens: **BARBADOS. St. Andrew:** On seashore, sandy dune area between beach and East Coast Road, between Belleplaine and Barclays Park, in full sun, forming a carpet, 21 Aug 1996, *Rogers 96-20* (BAR). **St. James:** On wave-splashed [seashore] rocks just S of Folkestone Park, 29 Oct 1996, *Rogers 96-126* (BAR). **St. Joseph:** Andromeda Botanic Gardens, 27 Jan 1998, *Rogers 98-4* (BAR). **St. Lucy:** seashore, on side of River Bay Inlet, saline, very wet, 19 Dec 1996, *Rogers 96-164* (BAR).

FIG. 66. *Zoysia matrella* (*Rogers* 96-164)
FIG. 67. *Zoysia matrella* (*Rogers* 98-4)

ACKNOWLEDGMENTS

The curators of the University of Michigan (MICH) and University of Florida herbaria (FLAS) allowed access to their collections during visits by G. Rogers. Dr. George Proctor, the only surviving co-author of the *Flora of Barbados,* visited Barbados and joined Rogers for field work. His rich experience has been a source of numerous insights. Dr. C. Dennis Adams through correspondence made a substantial contribution, especially in connection with *Paspalum.* Jill Marsden (K) identified *Dinebra retroflexa,* and a loan from the Royal Botanic Gardens Kew (K) helped with identifications in *Panicum*, *Themeda*, and *Zoysia.* Drs. Emmet Judziewicz and Paul Peterson made numerous suggestions for improving the manuscript. Additionally, the former provided tips on distinguishing *Olyra* from *Lithachne*, and the latter updated us on multiple nomenclatural matters. George Russell (US) sent information on *Themeda* material in the U.S. National Herbarium. Dr. Louis Chinnery and Dr. Angela Fields helped A. Holder learn laboratory techniques used in the *Paspalum* portion of the study. Dr. Sean Carrington collected *Setaria tenax* and brought it to our attention. Former UWI student Sadee Forde contributed information, ideas, and herbarium specimens from her 1996/97 research project on the Seashore Grasses of Barbados. A substantial portion of the bibliographic work took place in the Kent State University and University of Florida libraries. Our projects concerning Barbados grasses would have been impossible without generous financial support, and for this we thank the Peter Moores Foundation, Sam Lord's Castle [resort hotel], Roberts Manufacturing [Wayne Feeds], the Port St. Charles Development Company, and the University of the West Indies. Sophia Gremmel photographed the cover.

DOUBTFUL AND EXCLUDED SPECIES

Our account differs a little in scope from the *Flora of Barbados*, which included cultivated species. We have included strictly cultivated ornamental and household species in the keys to their genera but have denied them full treatments. Crop species were given no attention. The strictly cultivated species include *Bambusa glaucescens (B. multiplex* (Lour.) Rausch ex J. A. & J. H. Schult.), *Pennisetum setaceum* (Forssk.) Chiov., *Saccharum officinarum* L., *Sorghum bicolor* (L.) Moench. (as "*S. saccharatum* (L.) Moench." in Gooding et al., 1965), *Thysanolaena maxima* (Roxb.) Kuntze (see discussion of *Arundo*), and *Zea mays* L. Gooding et al. (1965) also listed the cultivated species *Dendrocalamus strictus* (Roxb.) Nees and *Tripsacum laxum* Nash, though we have no personal experience with these in Barbados. The two *Cymbopogon* species in the key are not known to be components of the uncultivated flora, but we gave *C. citratus* a full treatment because it is abundant in local gardens and may persist from prior cultivation in places.

Some uncultivated species reported for Barbados by Gooding et al. (1965) and by others are not well-substantiated. The "chain of evidence" for species treated formally or mentioned casually in the *Flora of Barbados* (and infrequently elsewhere) but not encountered in the 1996–99 field work varies in quality. Because the *Flora of Barbados* account comes primarily from an earlier work that was not based on preserved specimens, there is no "hard" evidence for some reported but missing species. These are:

Cenchrus brownii Roem & Schult. Listed by Gould (1979) for Barbados with no specimens cited. A great deal of checking has failed to yield this species in BAR or in the field.

Digitaria horizontalis Willd. Mentioned casually by Gooding et al. (1965) with no specimen citation, no specimen in BAR, and no locality recorded.

Oplismenus burmannii (Retz.) P. Beauv. Plausibly reported by Gooding et al. (1965) but no specimens cited nor known, not encountered 1996–1999 despite extensive checking, including checking the reported localities (where *O. hirtellus* was encountered).

Paspalum breve Chase. Casual mention in Gooding et al. (1965) with neither specimen citations nor localities. Probably not present in the Lesser Antilles.

Paspalum setaceum Michx. Possibly *P. laxum* misidentified though potentially collected once. See discussion of *P. laxum.*

REFERENCES

ADAMS, C.D. 1972. Flowering plants of Jamaica. Mona, Jamaica: Univ. West Indies.

ALLAN, J.A. 1957. The grasses of Barbados. London: Her Majesty's Stationery Office.

AMSHOFF, G.J.H. and T.H. HENRARD. 1943. Gramineae. In: A. Pulle, ed. Flora of Suriname. Vol. 1, Pt. 1. Leiden: E. J. Brill. Pp. 273–442.

BAAIJENS, G.J. and J.F. VELDKAMP. 1991. *Sporobolus* (Gramineae) in Malesia. Blumea 35:393–458.

BANKS, D.J. 1966. Taxonomy of *Paspalum setaceum* (Gramineae). Sida 2:269–284.

BOR, N.L. 1970. Gramineae. In: K. Rechinger, ed. Flora Iranica, Graz. [Not seen.]

CLAYTON, W.D. 1964. Studies in the Gramineae VI. Sporoboleae. The *Sporobolus indicus* complex. Kew Bull. 19:287–296.

———— and S. A. RENVOIZE. 1986. Genera Graminum. Grasses of the world. Kew Bull., Addit. Ser. 13.

CORRELL, D.S. and H.B. CORRELL. 1982. Flora of the Bahama Archipelago (including the Turks and Caicos Islands). Vaduz: Cramer.

DAVIDSE, G. and R.W. POHL. 1994. Poaceae. In: G. Davidse et al., eds. Flora Mesoamericana. Ciudad Universitaria, Mexico: Univ. Nac. Aut. Mex., Missouri Bot. Gard., & Nat. Hist. Mus. (London). 6:184–402.

FOURNET, J. 1978. Flore ilustree des phanerogames de Guadeloupe et de Martinique. Paris: Institut National de la Recherche Agronomique.

GOODING, E.G.B. 1974. The plant communities of Barbados. Bridgetown, Barbados: Government of Barbados.

————, A.R. LOVELESS, and G.R. PROCTOR. 1965. Flora of Barbados. London: Her Majesty's Stationery Office. [G. R. Proctor is author for grasses.]

GOULD, F. 1979. Poaceae. In: R.A. Howard, ed. Flora of the Lesser Antilles. Cambridge, MA: Harvard Univ. 3:25–220.

GREUTER, W., ET AL. 1994. International Code of Botanical Nomenclature (Tokyo Code). Regnum Veg. 131:1–389.

GUEDES, M. 1976. The case for *Paspalum distichum* and against futile name changes. Taxon 25:512–513.

HALL, D.W. 1978. The grasses of Florida. Ph.D. dissertation. Univ. of Florida, Gainesville.

HENRARD, J.T. 1950. Monograph of the genus *Digitaria*. Leiden: Univ. Leiden.

HITCHCOCK, A.S. 1936. Manual of the grasses of the West Indies. U.S.D.A. Misc. Publ. 243:1–439.

————. 1950. Manual of the grasses of the United States. 2nd edition. U.S.D.A. Misc. Publ. 200:1 –1051.

———— and A. CHASE. 1917. Grasses of the West Indies. Contr. U.S. Natl. Herbarium 18(7):261–471.

JUDZIEWICZ, E.J. 1991 [as 1990]. Poaceae. In: A.R.A. Gorts van Rijn, ed. Flora of the Guianas. Koenigstein: Koeltz Scientific Books. 8:1–727.

KARTESZ, J.T. and K.N. GANDHI. 1992. *Chloris barbata* Sw. and *C. elata* Desvaux (Poaceae), the earlier names for *C. inflata* Link and *C. dandeyana* Adams. Rhodora 94:135–140.

McVAUGH, R. 1983. Gramineae. In: Flora Novo-Galiciana 14:1–436.

POHL, R. 1980. Gramineae. In: W. Burger, ed. Flora Costaricensis. Fieldiana, Bot. New Series 4:3–608.

PROCTOR, G.R. 1984. Flora of the Cayman Islands. London: Her Majesty's Stationery Office.

ROGERS, G. 1998. The natural history of grasses and sedges in Barbados. Jour. Barbados. Mus. Hist. Soc. 44:72–106.

———— and A. HOLDER. 1998. The *Paspalum distichum-P. vaginatum* species pair on Barbados (Poaceae). Moscosoa 10: 47–56.

SAUGET, J.S. (HERMANO LEON) and E.E. LIOGIER (HERMANO ALAIN). 1974. Flora de Cuba. Vol. 1. First published in Havana 1946–1953, reprinted by Koeltz Scientific Books, Koenigstein.

VELDKAMP, J.F. 1973. A revision of *Digitaria* Haller (Gramineae) in Malesia. Blumea 21:1–80. [Not seen.]

————. 1990. The true identity of *Sporobolus poiretii* (Gramineae). Taxon 39:327–328.

VERNON, K.C. and D.M. CARROLL. 1965. Soil and land-use surveys. No. 18. Barbados. Trinidad: The Imperial College of Tropical Agriculture.

WATSON, L. and M.J. DALLWITZ. 1994. The grass genera of the world. Revised edition. Wallingford: CAB International.

WEBSTER, R.D. and P.M. PETERSON. 1996. Poaceae. In: P. Acevedo-Rodriguez, ed. Flora of St. John, U.S. Virgin Islands. Mem. New York Bot. Gard. 78:509–540.

INDEX

The accepted scientific names of plants of the Barbados flora are in [Roman type]. In addition, accepted generic names of plants in the flora are in [**bold**]. Taxonomic synonyms and names of plants casually mentioned are in [*italics*]. Common names are in [SMALL CAPS].